D0984937

GLOBAL MOBILE SATELLITE SYSTEMS

A Systems Overview

GLOBAL MOBILE SATELLITE SYSTEMS

A Systems Overview

Edited by

Peter A. Swan, Ph.D.
SouthWest Analytic Network
Paradise Valley, Arizona

Carrie L. Devieux Jr., Ph.D.
Chandler, Arizona

KLUWER ACADEMIC PUBLISHERS
Boston / Dordrecht / London

Distributors for North, Central and South America:
Kluwer Academic Publishers
101 Philip Drive
Assinippi Park
Norwell, Massachusetts 02061 USA
Telephone (781) 871-6600
Fax (781) 871-6528
E-Mail: < kluwer@wkap.com>

Distributors for all other countries:
Kluwer Academic Publishers Group
Post Office Box 322
3300 AH Dordrecht, THE NETHERLANDS
Telephone 31 78 6576 000
Fax 31 78 6576 254
E-Mail: < services@wkap.nl>

 Electronic Services < http://www.wkap.nl>

Library of Congress Cataloging-in-Publication Data

Global Mobile Satellite Systems *A Systems Overview*
Peter A. Swan and Carrie L. Devieux, Jr. (Eds.)
ISBN 1-4020-7384-4

Printed on acid-free paper.

Printed in the United States of America

Dedication

This book is dedicated to the Memory of Dr. Burton I. Edelson, a Pioneer in the field of Satellite Communications and former Director of COMSAT Laboratories, Clarksburg / Maryland, USA. His contributions were many in the world of Satellite Communications. His ideas flourish in the present, and his vision will certainly become reality

Contents

Foreword

The challenges that the embryonic Global Mobile Satellite Systems industry was facing in the early 1990's were overwhelming. The common impression was that it was improbable that the programs could meet their schedule because of the many revolutionary things that had to be accomplished.

-Build two hundred satellites! Who had done that?
-Launch two hundred satellites within a year! Can the launch industry change?
-Integrate software for orbital cellular infrastructure! Several Million lines of code!
-Operate total constellation from one site! Many satellites per person vs. the other way around!
-Handsets that can talk to satellites! Suitcase/Laptop size predicted in 1990.
-Maintain schedule with many critical paths! A mega-project accomplished on time?

The excitement inside the projects was phenomenal. The human commitment was remarkable with intellectual inventions, long nights, many days on the road, and concurrent engineering with designers, operators and launch professionals. The expectation within the projects was that they would meet the start of operations date with commercial viability. The dreams were large and the risks appropriate.
Indeed, the engineers, program managers, comptrollers, and schedulers

succeeded with on-time delivery of space constellations. This book is about the successes that occurred during the decade of the 1990's. It looks at the various approaches deemed reasonable for success in business with moderate risk. This is an interesting story, and one that needs to be told. The authors of this book put together a very good story line with the experience and successes laid out in the chapters. The rewarding aspect of the engineering successes is remarkable and self-evident.

We would like to add our thanks to the authors for placing the story in a manner that shows the trials of the engineer and the program participants. These challenges and the resulting lessons learned are valuable to those who will follow with similar Mega-Projects that "take on the world." We did enjoy the ride and look forward to the next decade of customers using the engineering accomplishments.

Preface

This book gives a systems overview of mobile satellite communications. The systems that are considered during this discussion are called Global Mobile Satellite Systems, or GMSS, and include active programs such as Globalstar, IRIDIUM®, Orbcomm, ACeS, and Thuraya. The book starts with a quick description of the three generations as they are being described in the telecommunications industry: First Generation – geosynchronous mobile satellite systems such as Inmarsat; Second Generation – GMSS systems; and, Third Generation – High data rate systems such as Inmarsat-Horizons, or New ICO. Selected trades are identified and explained and show how various GMSS systems were formulated. These trades are put in the context of space system architectures that deal with critical factors such as regulatory, financial, international, and, of course, engineering.

This book lays out Global Mobile Satellite Systems in the "big picture" sense with a top level overview leading to more detail in the various topics addressed in each chapter. The top level overview includes the examination of market demand, business trades, regulatory issues and technical considerations. More detail is then provided to ensure an understanding of major issues facing mobile satellite communications programs. Major issues are laid out in trade study style to provide easy access to the key information backed by references, tables, equations and cost / benefit analyses. Critical understanding arises when the key systems drivers are identified and laid out in such areas as orbits (trades between Low Earth Orbit - LEO, Medium Earth Orbit - MEO, and Geosynchronous Earth Orbit - GSO), frequency, protocols (time division multiple access vs. code division multiple access), projected customer bases (voice, messaging, or data), regulatory (regional or global – Big LEO or Little LEO), and engineering (processing transponder payload vs. bent-pipe transponder payload; small, lower altitude satellites vs. higher orbit, larger satellites).

The objective of this book is to make mobile satellite communications understandable and is directed toward decision makers, engineering managers, regulators, financiers, engineers, and technicians. There is a requirement in the communications industry to understand this dynamic arena, especially as many programs have gone through financial problems (customer base being critical to success). Not only will success rest upon the engineering brilliance of the program teams, regulatory breakthroughs in the telecommunications arena around the globe, and the management of "mega-projects" in a timely manner; but, also, in the customer acceptance of subscriber units and financial contracts for global service. This book is organized in eight chapters:

Chapter 1: Introduction to GMSS – The Global Mobile Satellite Systems deployment represents the recognition of the necessity for global mobility. Little LEOs, Big LEOs, and Super GSOs are the categories of this second generation of GMSS. This chapter places these systems within the context of terrestrial and satellite mobile concepts.

Chapter 2: GMSS Architectural Comparisons – Space Systems architecture shows similarities and differences of the three major categories of second generation GMSS systems. This space systems architectural process develops and evaluates systems; and, is characterized by three early steps: capture of needs and requirements; development of architectural concepts; evaluation and review. The GMSS systems are compared with these three early design steps as a way to "walk through" system development.

Chapter 3: Market Demand Considerations –Market trends for GMSS systems are evaluated, based upon an historical review of terrestrial cellular. The consumer services market growth, with the typical "S-Curve" that has periods of awareness, early adopters, rapid growth, saturation and obsolescence, is compared to the uptake of products such as various consumer products and satellite direct broadcast television services.

Chapter 4: Regulatory and Spectrum Considerations – Regulatory and spectrum considerations for satellite systems that provide communications to mobile satellite users can be divided into two types: Regional and Global. This chapter will outline what frequencies have been allocated for use to global mobile satellite service on an international and regional basis, the international/regional organizations embroiled with the regulatory aspects of these bands, and the regulatory barriers that have to be dealt with to ensure equitable access to these frequencies.

Chapter 5: Orbital Trades for GMSS Missions – The movement off the GSO orbit baseline of the first generation of GMSS architectures towards the new arena of LEO and MEO has been exciting and complex. Constellation trades require an understanding of the coverage, timing, visibility, beam coverage, and network connectivity vs. trade spaces available for each orbit type, such as altitude, number of planes of orbits and number of satellites inside each plane. These trades are shown from an architectural viewpoint with customer satisfaction being a major player in projected success.

Chapter 6: Propagation - Propagation effects have a significant impact upon the design, performance, Quality of Service (QoS), service cost and many other aspects of mobile satellite communications. Propagation impact is summarized qualitatively and quantitatively for various satellite geometry, vehicle motion, shadowing environment, modulation and access techniques as well as various mitigation techniques to improve quality of service and other performance measures.

Chapter 7: The Space Segment – A communication satellite system is structured around a Mission payload (characterized largely by its uplink sensitivity, downlink RF power, coverage, and connectivity) and the Spacecraft Bus (characterized by power generation, attitude control, structural integrity and thermal control). This chapter discusses these basic satellite payload components and the technology considerations and trade-offs within a system framework

Chapter 8: Summary – This chapter will pull together various themes from the chapters and lay out a view of the future for Global Mobile Satellite Systems

Peter A. Swan, Ph.D.
Co-Editor
SouthWest Analytic Network, Inc
Paradise Valley, Arizona

Carrie L. Devieux Jr. , Ph.D.
Co-Editor
Ocotillo / Chandler, Arizona

Chapter 1

Introduction To Global Mobile Satellite Systems

Peter A. Swan[1] and Carrie L. Devieux Jr.[2]
[1] *SouthWest Analytic Network, Inc,* [2] *Chandler, Arizona*

1.1   INTRODUCTION

History is full with examples of bad information delivered too late, to the wrong place. From Julius Caesar to Pearl Harbor, the course of history would have been different if the message had arrived on time, whatever the cost. The classical need is for:

"Communications Anytime, Anyplace"

Humanity advanced from communications systems using smoke signals to the telegraph in several millennia, while the progress from telegraph to interplanetary communications took roughly one hundred years. The development of radio communications (wireless) somewhat paralleled that of the wired infrastructure (Public Switched Telephone System, or PSTN). It took about 50 years from the initial transatlantic radio transmissions to the commercial introduction of communications satellites based on the pioneering innovation of Arthur C. Clarke (Clarke 1945). The development of wireless cellular communications infrastructure which made mobile communications popular (communications while on the move) has been on-going for the last 20 years. Communications satellites operating over the

Clarke orbit (the Geostationary Orbit, or GSO•) provided substantial improvement of service quality and reliability over the commonly used High-Frequency (HF) communications services. Early GSO communications satellites (Butrica, 1997) were regional, limited to large dishes, and used restricted beam patterns. During the 80's and early 90's, the First Generation (1G) of mobile satellite systems at GSO (Marisat, INMARSAT and American Mobile Satellite Company – AMSC, TMI (Canada), Australia's Mobilesat/Optus) became operational. They are characterized by conventional bent-pipe transparent transponders (no processing on-board), by rather large antenna beamwidths with user equipment on-board ships or in-vehicle. The limiting power and beamwidth characteristics prevented the use of small portable hand-held equipment.

During all of this refinement in communications infrastructures and devices, the need for communications anywhere at anytime was not fulfilled. These gaps puzzled many engineers until the right confluence of forces existed to develop a successful business environment. During the last part of the 80's, the following major developments and environments converged:

(a)   Small satellites became cost-effective and capable
(b)   High speed processing became possible on-board spacecraft
(c)   A robust commercial launch vehicle market emerged
(d)   Global telecommunications environment centered around de-
      regulation
(e)   Cellular telephones won widespread acceptance
(f)   Consumers demanded communications anytime, anywhere
(g)   Handsets became smaller and less expensive
(h)   Détente opened up the globe to commercial ventures
(i)   International investors were looking for global projects
(j)   Internet usage expanded our vision of global connectivity

These events came together around the 1988-1992 time period with an amazing result. An interesting proliferation of innovative Global Mobile Satellite System (GMSS) concepts was matched with an amazing combination of investors and partners. It seems that now, the technology would support the global need; and, the VISION could be realized.

GMSS Vision: Communications Anytime, Anyplace

• A GSO satellite operates in the equatorial orbital plane at an altitude of 35,786 km; moves around its orbit at the same rotational velocity as the earth and hence appears stationary to an observer on the surface of the earth. [It is often referred to as a GEO satellite]

1.2   GENERATIONS of GMSS

The first generation (1G) of GMSS was developed because of the need for reliable communications across the maritime environment. Shipping lane density and a tremendous growth of international trade during the 50's and 60's encouraged the leveraging of new satellite communications technology for mobile platforms. The initial experiments, using the L-Band frequencies, were performed by NASA in the 1974-1979 time frame with the Advanced Technology Satellite (ATS-6) GSO satellite[+]. These were the first handheld experiments made possible by the use of a large (10 m.) deployable antenna on ATS-6. COMSAT initiated the MARISAT system with a UHF package on the Navy GAPSAT satellite around 1976. MARISAT demonstrated the feasibility of ship to shore communications. The ship communicated over L-Band frequencies with the GSO MARISAT satellite from a fixed, relatively small, dish on-board the ship. The GSO satellite used a transparent transponder to translate the signal to C-band for reception by a large dish antenna (called shore earth station). Shore earth stations provided connections to the terrestrial telephone network, PSTN (Public Switched Telephone Network). In this fashion, reliable communications could be established, for the first time, between a ship at sea and land headquarters. [See Table 1.1 below, First Generation GMSS].

This system of the 1970's tied together the business environments of the shipping arena and commercial GSO satellite systems and technologies, only twelve years after the first GSO communications satellite – SYNCOM III was put into orbit. The next step in the maturation of first generation (1G) GMSS was the formation of the INMARSAT consortium and their satellites. This international consortium (semi-government structure) consisted of member companies which represented governments and provided funding and mission requirements. INMARSAT has since become privatized. The first international mobile satellite telephone system for maritime services used the Inmarsat-A type of terminals. This system employed analog modulation (FM) and allowed ships to call headquarters while at sea. This was especially important for maritime safety as well as commercial and personal business communications. INMARSAT has been orbiting various generations of satellites with increasingly more advanced capabilities (such as narrower spot beams, digital modulation), which allow communication with smaller user antennas (e.g. Inmarsat-C services). This led to portable

---

[+] In the 1980's, NASA also developed the Advanced Communications Technology Satellite (ACTS) to provide broadband FSS (fixed satellites services) in the 30 GHz band of the spectrum (Gedney, 2000). These services are for users at a fixed location as compared to GMSS users who can move to any location

briefcase-size, and later lap-top size, units called Inmarsat-M and Mini-M[#] . These services rely on spot beam technology and are primarily targeted to users on land. By contrast, the Inmarsat B service provides 64 Kbps to maritime users, but requires considerably larger user hardware. The COMSAT Mobile Communications of COMSAT Inc. used to provide services with the Mini-M under the Planet-1 name brand. (COMSAT Mobile has since been acquired by Telenor, Norway). These systems are portable; but must be in a fixed position to communicate properly with the flat-plate antenna pointing at the INMARSAT GSO satellite. In addition to voice, these systems can be used for data rates as high as 64 Kilobits per second (Inmarsat M4 service). This is much higher than the data rate of the second-generation systems represented by IRIDIUM, Globalstar, the original ICO and others with limited data rates (2.4 Kilobits / second to 9.6 Kbps). In a sense, Inmarsat-M4 using the Inmarsat-III satellite is a step closer to the third generation systems which will provide still higher data rates (e.g. 144 Kbps). The Inmarsat-IV (Inmarsat-Horizons) expected to be in service around 2004 will provide such data rates. The New ICO system will provide also good data rates.

Subsequently, various studies were performed to develop a similar technology for land-mobile communications (moving vehicles using a roof-top antenna). Such studies were done by NASA/JPL under the MSAT-X program. Such an approach has been developed and implemented by AMSC (U.S.), TMI (Canada), and OPTUS (Australia). These GSO systems have the goal of providing Land Mobile Communications over a much greater area than terrestrial Land Mobile Systems. The service is mobile; but, the phone could only be used inside the car. In other words, the service is not a PCS (Personal Communications System) as the telephone unit could not be carried anywhere by the user (telephone set not in the "handheld" personal mobility class).

The transition to the second generation (2G) of GMSS occurred during the latter part of the 1990's. The second generation is focused on connecting to the mobile user with personal "pocket-sized" units for messaging, voice, fax or data. The third generation (3G) is focused on "bandwidth on demand" to modest sized antennas for high-speed data and voice. The perception is that the explosion of the high speed data arena, such as global internet, is creating a need for this third generation of communications satellites for portable or mobile subscriber units. The GMSS growth pattern is shown in Table 1.2.

[#] It is important to distinguish between the names given to satellite constellations and those given the services / user terminals. The satellites are Marisat, ESA Marecs, Intelsat MCS, Inmarsat II, III, and IV (2004); the services / terminals are Inmarsat A, B, C, D, E, M.

Table 1.1 First Generation GMSS

TYPE	SYSTEM / SERVICE	YEAR
Analog- Maritime	NASA ATS-6 Marisat INMARSAT-A services	1974 1976 1982
Digital – Maritime	INMARSAT-C services	1988
Aeronautical	INMARSAT-M services	1993
Regional Land Mobile	Mobilesat / Optus AMSC TMI (Canada)	88/93 1992 1992

The description of the three generations of GMSS systems sets the stage for the remainder of the book. A large number of (3G) systems have been proposed in the form of filing of applications to the US Federal Communications Commission (FCC) and the International Telecommunications Union (ITU). It is not known at this time, how many of these systems will be developed. The market demand trends will certainly be a deciding factor.

Visions from the early days of GMSS, of maritime safety through GSO communications satellites, have matured to mobile communications for the individual and the concept of "Internet in the Sky." Table 1.2 shows many of these systems and compares such items as orbits and frequency bands.

1.3   GMSS CATEGORIES

Second generation GMSS solutions grew in the 1990's with diverse service objectives which lead to a few defined groupings, or categories. Table 1.3, GMSS Categories, shows the basic characteristics of each of three major categories. While GSO mobile satellite systems operated in the 90's, the business community invested also in the commercial LEO, MEO and SuperGSO* alternatives.

---

* SuperGSOs are characterized by the very large deployable satellite dishes needed to provide narrow spot beams on the ground. They operate at the geostationary orbit.

Table 1.2 GMSS Generations

	1st GENERATION	2nd GENERATION	3rd GENERATION
Features	Analog-Digital to Larger Antennas	Digital to Pocket units	High Speed Data
System	Marisat (76) INMARSAT A(82) INMARSAT C(88) AMSC (92)	Orbcomm (98) IRIDIUM (98) Globalstar (99) Ellipso ACeS (00) Thuraya (00)	New ICO INMARSAT Horizons 1997 FCC Filings [Thuraya (00)]
Frequency	Mobile L-band Maritime L-band Aeronautical	VHF / UHF L-band & S-band	L-Band & S-Band
General	GSO Regional Large User Antenna	LEO, MEO, GSO+ Small Handsets	LEO, MEO, GSO High Speed Data to Modest Antennas

GMSS categories were defined by the United States Federal Communications Commission (FCC) during the evaluation of submissions covering requests for frequency licenses. "Big LEOs" are basically systems providing voice services in the L and S frequency bands. "Little LEOs" are low data rate systems at UHF/VHF frequencies for data messaging. "Super GSOs" are regional mobile communications satellite systems operated from a 24-hour orbit. The term Big-LEO is often used in the broader sense to refer to non-Geosynchronous (non-Geosynchronous orbits, or NGSO). Specific systems, under the two GMSS LEO categories, may have an orbit that is low earth orbit, medium earth orbit, or even elliptical. This conflict in terminology originated from the need to name non-GSO systems in the FCC frequency licensing process. This resulted from the very large increase in the number of systems being proposed and the corresponding frequency licensing applications to the FCC. Table 1.4 LEO/MEO Constellations, illustrates the diversity in concepts, investors, owners, prime contractors , orbits and projected dates of operations for the plethora of systems filed through the International Telecommunications Union. It should be noted that

+ LEO: Low Earth Orbit; MEO: Medium Earth Orbit; GSO: Geostationary Orbit

this is shown for illustration only; a large number of these systems are not classified as GMSS.

Table 1.3 Major GMSS Categories

	Little LEO	**Big LEO**	**Super GSO**
Definition	Small Satellite, Low data rate, low cost systems	Complex Systems and Networks for voice, data and fax.	Regional systems for voice and data
Orbits	Lowest, inclined	LEO, MEO, polar, inclined, equatorial, and elliptical	Geosynchronous
Frequency	UHF and VHF	L – and S - Band	L - Band
Typical Systems	Orbcomm, Vita Sat, Leo One	IRIDIUM, Globalstar, New-ICO, Ellipso	Thuraya and ACeS

Figures 1.1, 1.2, and 1.3 on the forthcoming pages illustrate the systems architecture for each of the three GMSS categories. Little LEO is represented by the Orbcomm system; Big LEO is represented by the IRIDIUM system; while Super GSO is illustrated by the Thuraya system.

1.4 GMSS INFRASTRUCTURE

This section introduces the basic Global Mobile Satellite System (GMSS) infrastructure and compares it to the terrestrial cellular infrastructure. The extension of terrestrial systems to the arena of space helps explain some of the commonalities and shows significant differences between the two types of systems. The major advantage of satellite systems is that they provide coverage ubiquitously and can be effective in areas which would be inaccessible by any reasonably priced terrestrial systems. With a sufficient number of satellites, truly global coverage can be achieved everywhere for everyone.

Table 1.4. LEO/MEO Constellations (as of 1998)

LEO/MEO Constellations

Name	Prime	Owner	Type*/Freq	Planned Oper. Date	# Sats	Altitude Kms
Iridium	Motorola	Iridium, LLC	Big - L Band	98	66	780
Teledesic	-	Teledesic, LLC	Mega - Ka Band	2003	288	1375
Teledesic +	-	Teledesic, LLC	Mega - Ku Band	2005	30	10,320
Globalstar	Loral	Globalstar LP	Big - L/S Band	99	48	1414
Skybridge	Alcatel	Skybridge LP	Mega-KuBand	2002	80	1457
ECCO	Orbital Sciences	Constellation	Big - L Band		46	2000
Ellipso-Borealis	Boeing	Ellipsat(MCHI)	Big - L Band	2001	10	673x7515
Ellipso-Concordia	Boeing	Ellipsat(MCHI)	Big - L Band	2001	7	8060
?	Boeing	Boeing	Mega-KuBand		20	
Virgo		Virtual Geosat.	Mega-KuBand		15	
Signal	NPO Energia	KOSS	Big - L Band		48	1500
ICO	Hughes	ICO Global	Big -S Band	2001	10	10355
HughesNET	Hughes	Hughes	Mega - Ku Band		70	1490
HughesLink	Hughes	Hughes	Mega-KuBand		22	15000
WEST	MatraMarconi	MatraMarconi	Mega-KaBand		9	10000
Orbcomm	Orbital Sciences	Orbital Sciences	Little-VHF/UHF	99	21	785
Gonets D	AKO Polyot	Smolsat	Little-UHF		36	1400
Gonets R	AKO Polyot	Smolsat	Little-S/L Band		45	1400
Vitasat		VITA	Little-VHF/UHF	97	2	1000
FAIsat	Final Analysis	FACS	Little-VHF/UHF		26	1000
SAFIR	OHB System	OHB Teledata	Little-UHF		6	680
Temisat	Kayser-Threde	Telespazio	Little-UHF		0	938
Courier(Konvert)	NPO Elas	Elas Courier Com	Little-UHF	Cancelled		700
E-Sat		E-Sat(Echostar)	Little-VHF/UHF		6	1260
Iris(LLMS)	OHB System	SAIT Systems	Little-UHF		2	1000
Leo One		Leo One USA	Little-VHF		48	950
Eyetel	Interferometrics	GE Americom	Little		6	
GEMNet	CTA	CTA	Little		38	
KitCom	AeroAstro	KITCom	Little		6	
Marathon	NPO Prikladnoi	Marathon-Zemlya	Big		11	
Odyssey	TRW	TRW	Big	Cancelled		
Starnet	Alcatel Espace	Starsys Global	Little	Cancelled		

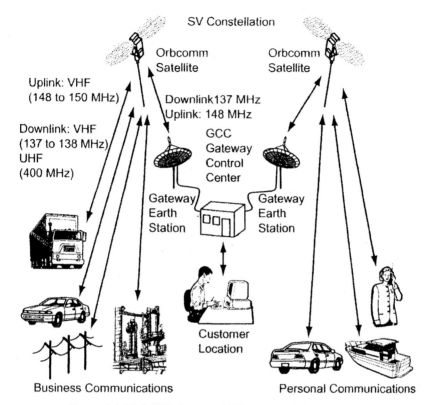

Figure 1.1 Little LEO System, ORBCOMM (Orbcomm WEB page1999)

1.4.1 Terrestrial Mobile Systems Infrastructure

The primary advantage of terrestrial cellular infrastructures is that they can handle millions of users simultaneously for high volume telephony. This means that cellular systems are designed for dense population areas resulting in infrastructures designed around small coverage areas called "cells." Therefore, wide area coverage cellular systems must have substantial infrastructures consisting of many cells. Their basic measure of coverage is in circular areas on the range of a few square miles. Historically, the first cellular systems of the 1980's were based upon an analog modulation system. . This technology still provides good voice quality over a reasonably "large" coverage area due to its radio characteristics (it takes advantage of the multipath propagation effects).

It was developed by AT&T and then implemented, tested, and deployed in the US, in Europe and Japan. Subsequently, various forms of digital transmission were introduced (time division multiple access – TDMA; code division code multiple access - CDMA). The following generations of

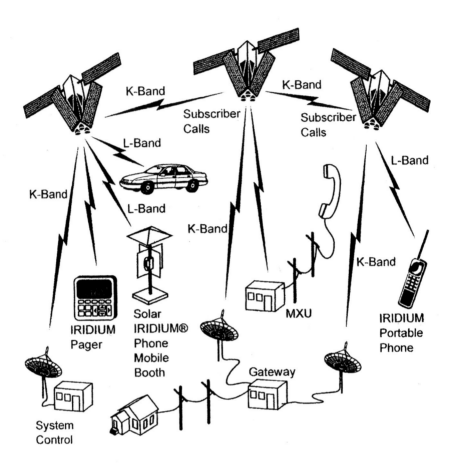

Figure 1.2 Big LEO System, **IRIDIUM** (Maine, Devieux, Swan,1995)

mobile telephony called personal communications systems (PCS),
employ digital TDMA or CDMA modulation, but operate over smaller
cells to improve capacity and quality of service. These systems have
various features allowed by digital telephony; and in particular, tend to
provide very good voice quality. Their geographical coverage is
limited to high density populated areas. One advantage of all terrestrial
systems is that they can increase system capacity by dividing cells into
smaller cells. Therefore, capacity can gradually increase to match
demand. It is not profitable for these terrestrial mobile systems to
cover areas, such as rural settings, where demand is low. This is where
mobile satellite systems can fill the gaps. GMSS can provide service in
the hard to reach and/or less developed areas of the world.

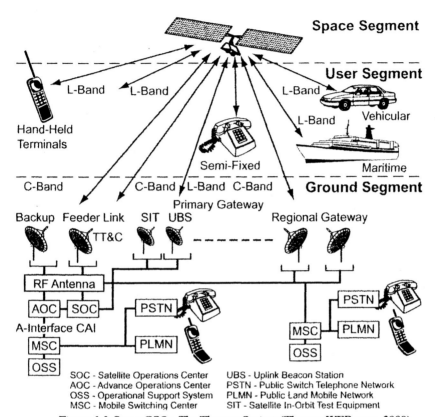

Figure 1.3 Super GSOs, The Thuraya System (Thuraya WEB page, 2000)

Moreover, GMSS have a superior advantage in emergency situations, such as in maintaining communications when all terrestrial power and telephony are disabled. GMSS, which allows direct communication from handset to handset (by-passing the terrestrial communications systems), such as the IRIDIUM system, are particularly well suited to these situations. The layering of various systems into a communications architecture is shown in Figure 1.4.

In simple terms, the communication link is between the cellular phone (handset) and a tower (called a base station) in a terrestrial cellular infrastructure. The tower is usually a fairly high structure for conventional cellular. It enables communications ranges of up to 10-20 miles. (This measure of distance is the definition of a cell. By comparison, the cell of an GMSS satellite beam is on the order of hundreds of miles in diameter.) As the user moves from one cell to the next, a handoff procedure switches the communications link into the next cell in a seamless manner.

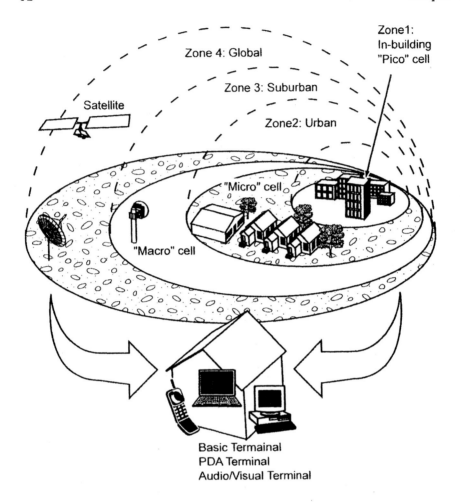

Figure: 1.4 Cell Coverage of Satellite, Analogue Cellular, PCS, and In-Building

A base station receives the signal from the user, processes it, and provides a duplex connection to the existing terrestrial telephone network or PSTN (public switched telephone network). The mobile customer is now connected to the desired local/international telephone system. Personal communications systems (PCS) operate with smaller cells and unobtrusive base stations that can easily be hidden on church spires or rooftops to meet neighborhood standards. These systems tend to rely on line-of-sight transmission. Another complication is the presence of physical blockage within the terrestrial network of PCS systems. Distances from user to tower must be smaller to adapt to these requirements with clear line of sight. The advantages are the superior clarity and high numbers of customers supported.

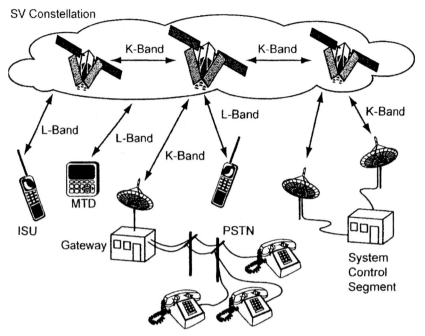

Figure 1.5 GMSS Infrastructure (**IRIDIUM**)

1.4.2   Global Mobile Satellite Systems

A typical GMSS infrastructure is depicted in Figure 1.5, above. Two similarities between the two cases of GMSS and terrestrial cellular are: (a) each satellite is a base station and is only one part of the mobile telephony network; and (b) the GMSS terrestrial gateway ties the orbiting base stations to the ground infrastructure similar to cellular base stations. A GMSS system gateway connects, via terrestrial lines, to a mobile telephone switching office (MTS) which connects to the PSTN. The IRIDIUM example shows how the satellites "base-stations" connect to gateways and also to the subscriber. This multiple satellite network enables ubiquitous coverage, even over the ocean. IRIDIUM has the advantage of using inter-satellite links which could allow world-wide coverage with only one gateway.

1.4.3   Summary Comparison of Major Features

One difference between the terrestrial and space cellular systems is the distance from user to base station; satellites measure in hundreds of kilometers for GMSS, while terrestrial systems measure in meters.

Table 1.5 Comparison of Systems

CLASS	Subclass	Quality	Handoff	Delay	Range
Terrestrial Cellular	Analog or Digital	Good	Not Frequent	Very Low	miles
	PCS, small cells	Superior	Most Frequent	Very Low	1000 ft
Satellite (MSS)	GSO	Good	Rare	High	20,000 miles
	NGSO / MEO	Good	Moderate	Low	6000 miles
	NGSO / LEO	Good	Frequent	Low	800 miles

Table 1.5 compares the various systems. An important difference between GMSS and terrestrial cellular is the delay (latency) experienced during a conversation. The significant distance the radio waves have to travel from user to satellite and back causes this delay. GSO systems can have one-way delays as high as 0.25 second while lower delays for Non-GSO (NGSO) orbits are due to their closer ranges. The International Telecommunications Union recommends that one-way delay for voice be limited to 0.3 second as conversations become difficult beyond this limit. Multipath radio propagation effects for GMSS although similar to terrestrial systems, deserve special consideration. An example of the received signal waveform measured at a mobile radio as well as other results can be found in (Devieux, 1993).

With significant differences between all the terrestrial (Analog, TDMA, CDMA, PCS) and GMSS (GSO, LEO, MEO) systems, the logical solution for the customer is to use the strength of each system in a complementary manner. As such, the use of dual-mode and tri-mode phones allows the optimal use of all systems types. A dual mode phone is one that can switch between two telephony systems, such as a terrestrial GSM system and a satellite GMSS system. Another example would be a Globalstar phone handling a CDMA digital system connection, when available; and the Globalstar CDMA space system connections as an alternate, when available. This would indeed enable the era of global roaming communications.

1.5 BOOK ORGANIZATION

This book is laid out as a top-level overview of the second generation of Global Mobile Satellite Systems with special emphasis on options and approaches that determine designs and architectures. It also addresses technical individuals and engineers. The basic content of this book focuses on the second generation and is aimed at engineering managers, regulators, financiers, communications engineers, and others interested in the basic elements of GMSS. The following is a summation of each chapter.

Chapter 1 - Introduction to GMSS: Global Mobile Satellite Systems are a product of the last decade of the 20th century. This peak in the telecommunications revolution represents recognition of the global mobility needs. This breakout of satellite systems is structured around the second of three GMSS generations. Little LEOs, Big LEOs, and Super GSOs are the categories of this second generation. This chapter places these systems within the context of terrestrial cellular and satellite mobile concepts.

Chapter 2 - GMSS Architectural Comparisons: Three major categories of the second generation of GMSS systems are compared as a method of studying various architectures. This shows similarities and differences of categories and individual systems. The space systems architectural process develops and evaluates systems. It is characterized by three early steps: capture needs and requirements; develop architectural concepts; evaluation and review. These processes are our foci as GMSS are developed from global needs toward operational networks.

Chapter 3 - Market Demand Considerations: Market trends for GMSS systems are examined. Much of the comparison is based upon a historical look at terrestrial cellular and services with the typical "S-Curve" that have periods of awareness, early adopters, rapid growth, saturation and obsolescence. Also presented as an example is the uptake of products and services such as VCRs and satellite direct broadcast television.

Chapter 4 - Regulatory and Spectrum Considerations: Regulatory and spectrum considerations for satellite systems that provide communications to the mobile satellite user can be divided into two types of systems; regional and global. Regional satellite systems, typically GSO satellites, provide coverage in a few regions of the Earth. The primary regulatory considerations are local requirements that require co-ordination of frequencies with other regional systems to guarantee satisfactory regulatory conditions exist to ensure effective operations. Global communications systems, typically provided by some type of non-GSO satellite network, require regulations and co-ordination on both a local and global scale. This chapter will outline what frequencies have been allocated for use to mobile satellite services on an international and regional basis, the international/ regional organizations involved in the regulatory aspects, and the regulatory barriers that have to be dealt with to ensure access to appropriate frequencies.

Chapter 5 - Orbital Trades for GMSS Missions: The movement from the first generation of GMSS architectures, based upon the GSO orbit, to the new arena of LEO and MEO, for the second generation, is exciting and complex. Constellation trades require an understanding of the coverage, timing, visibility, beam coverage, and network connectivity vs. allocations available for each orbit type, such as altitude, number of planes of orbits, and number of satellites inside each plane. These trades are shown from an architectural viewpoint with customer satisfaction a major player in projected success.

Chapter 6 – Propagation: Propagation effects have a significant impact upon the design, performance, Quality of Service (QoS), service cost and many other aspects of mobile satellite PCs communications. A substantial body of knowledge, as well as many proven mathematical models, exists for terrestrial cellular. Considerable progress has been made in understanding and modeling mobile satellite fading environments. Additional work is needed in many areas. Propagation impact is summarized qualitatively and quantitatively for various types of satellite geometry, vehicle motion, shadowing environment, modulation / coding / access techniques, as well as various mitigation techniques to improve Quality of Service and other performance metrics. Considerations related to future, third generation systems, are summarized.

Chapter 7 – The Space Segment: A communication satellite payload is characterized largely by its uplink sensitivity, downlink RF power, coverage, and connectivity. The spacecraft bus subsystem is characterized by thermal, structural, propulsion, and power requirements. This chapter discusses these basic satellite payload components as well as technological considerations and trade-offs within a system framework. Differences between LEO, MEO and GEO satellite payloads and specifications are reviewed.

Chapter 8 – Summary: This chapter puts together the various themes of the chapters and lays out a comprehensive look at Global Mobile Satellite Systems.

1.6 THE CHALLENGE

This book presents Global Mobile Satellite Systems in a manner that shows how each of the business teams accepted the challenge. The last two decades have provided a rare environment where engineers could significantly change the world with successful projects. The dream of having communications anywhere at anytime for anyone is beginning to be fulfilled. Not only will the players feel a satisfaction for finishing a worthwhile engineering project, but they will have the realization that their ten year projects will be beneficial to millions of people and significantly improve lives. Table 1.6 shows the challenges for the future while illustrating the successes of the past and present.

Table 1.6 GMSS Systems by Generation

Generation	System	Type of SV (Space Vehicle)	# SVs	Altitude (Km)	Status
FIRST (1st G)	INMARSAT	MSS	4	GSO	Operational
	AMSC / TMI	MSS	2	GSO	Operational
	MobileSat (Optus)	-		GSO	Operational
SECOND (2nd G)	Leo One	Small-VHF	48	950 Km.	Licensed 98
	Orbcomm	Little-VHF/UHF	28	785	Licensed 94 Operational
	Vitasat	Little-VHF/UHF	2	1000	Licensed 95
	FAIsat	Little-VHF/UHF	26	1000	Licensed 98
	E-Sat	Little-VHF/UHF	6	1260	Licensed 98
	IRIDIUM	Big - L Band	66	780	Licensed 95 Operational
	ECCO	Big - L Band	46	2000	Licensed 97
	Ellipso-B	Big - L Band	10	673 x7515	Licensed 97
	Ellipso-C	Big - L Band	7	8060	Licensed 97
	Globalstar	Big - L/S	48	1414	Licensed 95 Operational
	ICO	Big -S Band	10	10355	
	ACeS	Super GSOs	2	GSO	Operational
	Thuraya	Super GSOs	2	GSO	Operational
THIRD (3rd G)	INMARSAT-4	L-Band		GSO	
	New ICO	S-Band	12	10355	

REFERENCES

BOOKS

Rechtin, Eberhardt and Mark W. Maier. *The Art of Systems Architecting*. 1997. Boca Raton CRC Press.

Rechtin, Eberhart. *Systems Architecting, Creating and Building Complex Systems*. 1991, Englewood Cliffs, New Jersey: Prentice Hall.

REPORTS

Armbruster, Pete and Laurin, Mala " The IRIDIUM® Network for Global Personal Communications" Telecommunications Review, vol.6, No.6, 1996 Korea Mobile Telecom.

Butrica, Andre J. Editor " Beyond The Ionosphere- *Fifty Years of Satellite Communications"* The NASA History Series 1997.

Ellipso, The Affordable Connection. Marketing Package, 1999. Ellipso, Inc.

Globalstar Communications Magazine. Vol. 7, April/May 1999.

Orbcomm, Global Data & Messaging. Marketing Package, 1999. Orbcomm USA.

State of the Space Industry, 1997 Outlook. Reston, Virginia: Space Publications.

State of the Space Industry, 2000 Outlook. Arlington, Virginia: International Space Business Council.

JOURNAL/SYMPOSIA ARTICLES

Clarke, Arthur C. "Extra Terrestrial Relays: Can Rocket Stations Give World-Wide Coverage ? " Wireless World. October 1945.

Leopold, Raymond J. and Ann Miller. "The IRIDIUM Communications System." 1993 (paper # IEEE 0278-6648). IEEE Potentials. April 1993.

Maine, Kris; Devieux Carrie L. Jr.; and. Swan, Peter A. "Overview of IRIDIUM® Satellite Network," IEEE Western Communications Satellite Systems Conference, San Francisco, California, November 1995.

Swan, Peter A. "Space Systems Architecture: A Constellation Necessity," (IAF-97-M.3.01), 48th Congress of the International Astronautical Federation, Turin, Italy, 6 October 1997.

Taylor, Stuart C. and A. R. Adiwoso. "The Asia Cellular Satellite System." 1996. (paper # AIAA-96-1134-CP). American Institute of Aeronautics and Astronautics, Inc.

Devieux, Carrie L. Jr.. " Systems Implications of L-Band Fade Data Statistics for LEO Mobile Systems" International Mobile Satellite Conference pp. 367-372 NASA / JPL 1993

Chapter 2

GMSS Architectural Comparisons

Peter A. Swan
SouthWest Analytic Network, Inc.

2.1 INTRODUCTION

Understanding various constellations of satellites in the second generation of Global Mobile Satellite Systems (GMSS) requires an insight into the many choices made throughout their development. Three major categories of the second generation of Global Mobile Satellite Systems (GMSS) are compared as a method of studying the various choices. These systems similarities and differences (**Little LEOs**, **Big LEOs**, and **Super GSOs**) are quite indicative of the development process and trade studies conducted. Recognition of actual systems choices facilitates an understanding of the space systems architectural design process. This engineering design process is multifaceted and follows a space systems architectural approach developed by Dr. Eberhardt Rechtin of the Aerospace Corporation and the University of Southern California [Rechtin, 1991, 1997]. Insight gained from comparing the three major categories of GMSS strengthens our understanding of the similarities and differences between these systems. As space systems become more complex, and have multiple customers and stakeholders, a broader systems architectural view is mandated at the beginning of the process. Merging of critical desires into an engineering requirements document must be supported throughout the design. A combination of a definitized engineering process, artistic vision, and input of "soft" requirements yields a powerful step-by-step design process for development of a communications satellite system network. This process is called space systems architecture. The process steps within space systems architecture are our foci in this chapter as Global Mobile Satellite Systems are developed, from global needs towards operational networks. This chapter will initially define the space systems architecture process, and then use example systems to refine our understanding of this process.

2.2   ARCHITECTURAL DESIGN PROCESS

This section develops a view of multiple satellite constellation architectures and discusses various trades inherent in complex global networks in space. To provide this insight, the process of architectural development and design will be presented with example cases based upon Global Mobile Satellite Systems (GMSS).

2.2.1   Space Systems Architecture

Space Systems Architecture is a relatively new phenomenon when evaluated in historical perspective. Dr. Rechtin, in his book <u>Systems Architecting</u> [Rechtin, 1991], recognized that over time "...great architectures required creative individuals capable of understanding and resolving problems of almost overwhelming complexity." As a result... "Architecting... [has become] both a science and an art. The former is analysis-based, factual, logical, and deductive. The latter is synthesis-based, intuitive, judgmental, and inductive. Both are essential if modern systems architecting is to be complete." [Rechtin, 1991] Current challenges in developing large complex systems, such as the new space communications constellation systems with over 40 satellites, requires both of these skills. Especially important is artistic talent to understand customer desires and engineering skills to meld these within current technological feasibility. The tendency is to over promise toward the customer's future requirements and/or to over engineer towards an elegant, but delayed, technical approach.

Space systems architects deal directly with the customer and ensure a translation of needs to the systems engineering team that structures the design, development and manufacturing processes.   Architecting, as described by Webster, is *"the art and science of designing and building a system."* This definition explains that the systems architect, and especially the space systems architect, is routinely working in the realm of the artist and also in the discipline of the engineer. Thomas McKendree states a "...systems architect is a person who creates the conceptual model of the system, translating the client's desires into a technical description the builder can understand. The systems architect, as an agent of the client, must also ensure that system integrity is maintained throughout the program phases, and assures that design certification is meaningful and passable." [Rechtin, 1991]

2.2.2   The Space Systems Architectural Process

The job of a space systems architect is to create a space systems architecture that meets the needs, requirements, and constraints of the client. Emphasis for the architect, during the systems architecture process, encompasses three activities and reaches across the full spectrum of customers' needs. Figure 2.1 shows the system architecture interactive process as it spans development of the entire system. Systems architectural skills are especially critical during these tasks that make up the creative triangle. This upper left hand corner of the figure ties together the innovative phase with realities of design and production. These three creative steps, when the formative issues are addressed, are the essence of systems design. During comparison of the three categories of GMSS, this significance will become evident. Customer needs are compared with the manufacturability and launch costs during the creative process of architectural trade studies.

2.2.2.1   Capture Needs and Requirements
Development of requirements documents for the engineering design process is complex and multi-level. Development of relationship with the client is critical to successful identification of needs, requirements, and desires of customers and clients. Multiple stakeholders of future systems have different needs and must be handled with care to ensure an excellent relationship throughout the development cycle. One key heuristic to remember during this time consuming process is that the system architect should not "...assume that the original statement of the problem is necessarily the best, or even the right, one."[Rechtin and Maier, pp26, 1997.] An iterative, time consuming and focused activity is necessary to refine desires of clients into design requirements. This requirements review process must be accomplished with the customer and result in a Customer Requirements Document (CRD) formalized at a full-up Systems Requirements Review (SRR).

2.2.2.2   Develop Architectural Concept
Development of a systems concept from the users' needs, or customers' ideas, is always a challenge. All engineers recognize the complexity of the systems development process as outside issues come into the process and are seen as disruptive to design activities.

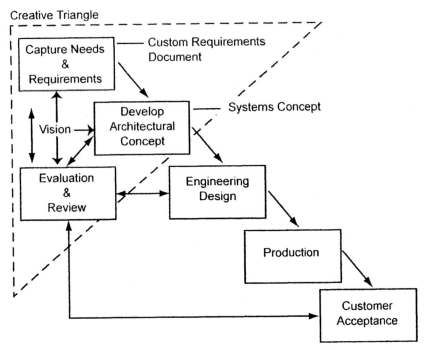

Figure 2.1: Systems Architecture Interactive Process

However, the systems architect must include all of the art aspects of the design process such as political, financial, corporate goals, resources, regulatory, and legal constraints. In addition, systems architects must incorporate their years of experience into the design process with a mixture of all the engineering disciplines leading to a concept that can be built within the constraints that have been identified. The key is that a space systems architect must start the complex iterative engineering design process with an intuitive conceptual development of a system. The first strawman of the final design must be articulated to stimulate and coalesce the process. No one expects this first concept to resemble the final product that is built; but a starting point is essential for the team and is a reference point from where to compare alternative designs. Four key factors for comparison are cost, risk, performance and schedule.

2.2.2.3 Evaluation and Review

This step in the process of systems architecture is an important iterative activity that matures early concepts into refined designs that can succeed. This process step is critical to eventual success as the systems architect is working in an unknown arena during the early phases and with design and manufacturing engineers during the later phases of the program. Trade studies are tools of systems engineers as issues are identified and refined. Evaluation and review by a systems architect includes key needs of both clients, such as international taboos, and those of design engineers, such as size, weight, and power. Through this evaluation and review process, a systems architect codifies and focuses the vision. Beyond identification of concept and needs of customers, a systems architect becomes a facilitator of simplicity during each phase of a developmental program. Continual evaluation of the progress and needs of customers should ensure maintenance of system integrity and reduction of complexity. This involvement can assure the clients during the latter phases of the program that the product will satisfy their needs and desires. Continued reference to original needs statement, vision, and systems requirements documents are essential for space systems architects during this phase. These activities are described further in Table 2.1, Architectural Steps.

2.3 GMSS ARCHITECTURAL STEPS

This section explains three principal architectural steps of the creative triangle with one example from each of the GMSS categories. Iteration activity to refine requirements, customer "buy-in," elimination of inappropriate ideas, and engineering feasibility checks are not shown. Numerous trades are identified that must be conducted before final engineering design concept can be frozen for manufacturing. These trade studies span the spectrum of factors from business needs to engineering feasibility and regulatory issues to customer acceptance perceptions. Issues vary greatly across GMSS systems, such as how to control the satellites, control and optimise the network, manufacture satellites, obtain appropriate launch services, and choose orbits and numbers of satellites.

2.3.1 Capture Needs and Requirements

This process step focuses on the first column of Table 2.1, Architectural Steps, while it progresses through the architectural design process using three major categories of GMSS as design points for comparison.

2.3.1.1 Identify Stakeholders

This is key to future development as acceptance of concepts, requirements and architectures must be agreed upon for the process to move forward. Breadth of stakeholders and their issues includes anyone effected by the product or mission. While many critical members of this group are visible, a space systems architect must speak for the silent stakeholders or clients during the developmental process. (see Table 2.2, Stakeholders)

Table 2.1. Architectural Steps[1]

Capture Needs and Requirements	Develop Architecture Concept	Evaluation and Review
Identify, customers, stakeholders.	Brainstorm	Review Requirements Specifications
Capture and Develop client's requirements and needs	Develop prototypes and perform simulations	Review System design specification
Perform req's analysis	Explore alternate concepts	Perform checklist evaluation
Define Problem	Select concept (first strawman)	Track and document change
Capture relevant rationale and decisions	Apply appropriate heuristics	Track and document problem reports
Apply appropriate heuristics	Determine interfaces between system elements	Develop, Refine and Explain a Project Vision.
Specify the requirements in a Customer Requirements Document	Allocate requirements to system architecture	Reduce Complexity
Design for Manufacturability requirements development	Identify systems drivers	Maintain System Integrity
Determine system context and interfaces	Specify the architecture in a Systems Concept Document	
Conduct Systems Requirement Review	Capture relevant rationale/decisions	
	Specify the system interfaces in a document	

[1] Stone, Arthur G. III, "Architecting a Systems Engineering Process," pg 68 (adapted from).

Table 2.2. Stakeholders

Little LEO	Big LEO	Super GSO
Company Owners	Company Owners	Company Owners
Space Systems Engineers	Space Systems Engineers	Space Systems Engineers
Marketers	Marketers	Marketers
Disaster Relief Personnel	Disaster Relief Personnel	Disaster Relief Personnel
Stockholders	Stockholders	Stockholders
Major Shipping Owners	Wireless Telephony Giants	Regional Investors
	Handset Developers	Regional Telephony Transporters
	Global Investors	Regional cellular owners
	Global Roamers	Mobile Public in footprint

2.3.1.2   Definition of Problem

At the beginning of any project, there are initial concepts and ideas. Recording these, along with a description of problems being addressed, enables the process to begin.

Table 2.3. Problem Definition

Little LEO	Big LEO	Super GSO
Global Reach-short messages	Telephone from anywhere- anytime	Mobile anywhere at anytime
Cheap messaging	Page anywhere at anytime	Beam Pattern
Autonomous Information movement	Fax and computer data anywhere at anytime	Reliable signal levels around the world
Lowest Cost	Significantly expensive system	Minimum systems expenses
	High customer fees	Minimum ground complexity
	Satellite production	Competitive subscribers
	Launch Campaign	Lower Service Quality For lower rates

2.3.1.3   Identify Mission Statements

Mission statements are usually associated with the problems and perceived solutions. Refinement and iteration occur often in the early stages of a space systems design and development. Small changes in the mission statement sometimes result in significant impacts to cost, performance and/or schedule.

Table 2.4. Mission Statements

Little LEO	Big LEO	Super GSO
Build an inexpensive, store and forward messaging system	Build a global handheld, mobile, voice communications system	Build a regional mobile voice system with handheld subscriber units

2.3.1.4 Identify Mission Objectives

Mission statements can be broken into objectives that can be ranked by value added and traded during the design process. A mission objective is a quantifiable statement dealing with identifiable and measurable achievements.

Table 2.5. Mission Objectives

Little LEO	Big LEO	Super GSO
Enable low data rate messaging globally	Maintain premier position in mobile communications	Develop premier position in regional mobile connectivity
Small, inexpensive satellites	Ensure a profitable project	Enable anywhere in beam pattern
Create funding approach to be first to market	Offer phone and pager service	Create regional partnerships ensuring first to market
Gain global frequency allocation	Create a funding approach reflecting risk	Develop funding profiles allowing rapid full development
Phased approach at operations, starting in 97	Establish a partnership for production	Ensure frequency allocation around globe
Gracefully fill constellation for levels of performance	Establish a realizable path for frequency allocation	
	Begin operations in 98 for a fixed price of $3.4B	

2.3.1.5 Identify Initial Concepts

Development of the initial concept is essential to the progress of the program because it forces players to concentrate on architectural issues. This focus on development of an initial concept inevitably results in many factors surfacing as needs and requirements. Items such as ability to interface with launch vehicles, to co-ordinate with international protocols, and to operate in an environment that is hostile are evaluated when initial concepts are surfaced.

Table 2.6. Initial Concepts

Little LEO	Big LEO	Super GSO
Simple satellites	Instantaneous connection globally	Reliable data rate regionally
Lowest Earth orbit	Recognizable voice quality	Voice and data on the move
Cheap rockets to orbit	Low Earth Orbit	Medium apertures for the subscribers
Time delay acceptable	Interconnected crosslinks	Bent-Pipe Transponder satellites
Early operations capability	Processing satellites	
Very cheap system	Modest number of telephony gateways	Regional network connectivity through gateways
Remote operations	Handheld telephones and small pagers	Compete with cellular
Enabling global "tagging"	First to market	First to market
Very low price for user equipment	"Complementary to Cellular"	Leverage GSO space heritage
Less than millions of customers	1-2 million customers	Less than millions of customers
	Compatible with global wireless networks	Compatible with wireless networks regional

2.3.1.6 Define Customer Needs and Requirements

At this point in the design process, needs of the stakeholders must be written down with quantifiable requirements initiated at the A-level, or systems requirements level.

Table 2.7. Customer Needs and Requirements

Little LEO	Big LEO	Super GSO
Modest Return on Investment	Big Return on Investment	Big Return on Investment
Corporate success	Partnership success	Partner success
Technical credibility	Technical superiority	Technical credibility
Operations in 97	Operations in 98	Operations in 2000
Thirty minute timeliness	Compatible with PSTN's	Compatible with future cellular
Five year lifetime	Five year lifetime	15 year lifetime
Natural environment	Natural environment	GEO environment
Low BER demand	Modest BER	Moderate BER demands
	Availability of 98.5 %	Availability of > 99 %

2.3.2   Develop Architecture Concept

This step is inside the creative triangle because imagination and technical experience are required to realistically feed back into the formative process. A significant process step, listed under Evaluation and Review process is the creation of a vision. To succeed in a major developmental program, a vision must be established and bought-into by all participants. It is key that everyone working on the project knows the critical factors for success and supports actions to achieve them. If identification of a vision occurs early in project, the team must embrace it and articulate it across the partners and players.   Once the feedback and vision is initiated, process steps are continuous and iterative. Updating specifications, trading documentation and changes, and reducing complexity requires diligence and perseverance. There are many actions during this step in the architectural process that includes items in the second column of Table 2.8. Some of the following expansions illustrate similarities across the three categories of GMSS systems.

Table 2.8. Architectural Steps

Capture Needs and Requirements	Develop Architecture Concept	Evaluation and Review
Identify customers, & stakeholders.	Brainstorm	Review Requirements Specifications
Capture and Develop client's requirements and needs	Develop prototypes and perform simulations	Review System design specification
Perform requirements analysis	Explore alternate concepts	Perform checklist evaluation for completeness
Define Problem	Select concept (first strawman)	Track and document change requests
Capture relevant rationale and decisions	Apply appropriate heuristics	Track and document problem reports
Apply appropriate heuristics	Determine interfaces between system elements	Develop, Refine and Explain a Project Vision.
Specify the requirements in a Customer Requirements Document	Allocate requirements to system architecture	Reduce Complexity
Design for Manufacturability requirements development	Identify systems drivers	Maintain System Integrity
Determine system context and interfaces	Specify the architecture in a Systems Concept Document	
Conduct Systems Requirement Review	Capture relevant rationale and decisions	
	Specify the system interfaces in a document	

2.3.2.1   Brainstorm

Key to successful development of a system in a timely and cost-effective manner, is the ability to create during a dynamic period of developing and changing requirements. This capability, to forge an initial concept early in the process, when almost everything is unknown or has very dynamic requirements, demands discipline to force structure when very little exists. As a result, initial efforts almost always lead to concepts that compete against each other. A productive method to identify these alternative concepts is use of brainstorming (the process of thinking freely without constraints). This creative activity should reward ideas; good, bad, or silly. The repetitive approach (to beat requirements against concept feasibility) is the iterative process that usually produces the best results. The statement "First Strawman on the Street - WINS" is reflective of original brainstorming efforts; and, usually results in major features surviving the iterative design process and establishing the reality of the project. The first strawman may change as alternative solutions are developed and compared; or, it many remain largely intact. Each iteration after the first strawman contributes to the final design.

Table 2.9. Brainstorming Ideas

Little LEO	Big LEO	Super GSO
Balloon, Aircraft	Balloon, Aircraft	GEO heritage
GEO, MEO, LEO	GEO, MEO, LEO, HEO	Smaller and more rapid deployment
Polar, elliptical, equatorial	Polar, elliptical, equatorial	One satellite with backup
Real time or delay	Crosslinks or ground	Wide area coverage
Store and forward	Processing or Bent-Pipe	Simplicity and Bent-Pipe
Smallest subscriber units	Large ground relay infrastructure	Initially, car telephone size
Tagging everything	Large ground antennas and small satellites	Simple ground infrastructure

2.3.2.2   Explore Alternative Concepts

During this step in the creative triangle of the systems architect, several major actions must be accomplished. Many can be done in parallel, or even out of sequence, as this period of development is extremely uncertain. Steps outlined in Table 2.8 will add structure to the developmental process and assist in the refinement of a preliminary mission concept. Such processes as brainstorming, capturing reverent rationale and decisions seem to be counter in approach and style. Indeed, they are; and, to be accomplished well, they require people with adaptive skills. Development of architectural concepts requires a diverse team with talents ranging across the engineering and management spectra.

The command for this design process is: create alternative concepts and systems architectures. This is a critical process with a responsibility to ensure complete coverage. If the mission is to communicate over a certain region, this process should ensure that towers, balloons, and aircraft are considered as well as spacecraft. Alternative concepts should be encouraged, at this phase of the program, as a way of ensuring coverage of all concepts. Despite schedule pressures, the final strawman design concept should never be allowed to be institutionalized as the solution until all alternates are traded against all requirements. Institutional biases, such as a strength/history in LEO or GSO systems, are surprisingly strong and often prevent a complete assessment of all alternatives. Stimulation of alternative concepts surfaces with surprising by-products. Creative products are very difficult to mandate; but, they can be stimulated with an environment where reward is visible and ridicule is non-existent. During this process, optimum solution terminology should not be used. Important factors can emerge along with the identification of major segments of the concept. [Table 2.10].

Table 2.10. Alternative Concepts

Little LEO	Big LEO	Super GSO
One satellite or multiple	Large satellites or small	Simple satellites at GEO
GEO vs. LEO	Processing or Bent-Pipe Transponder	Large antenna/small subscriber units
Large ground antennas or hand held devices	Crosslinks or ground infrastructures	Simple Transponder payloads
	Aircraft relays	Direct to ground network
	MEO vs. LEO vs. GEO	One satellite multiple beams
	Polar vs. elliptical vs. inclined	Start regional, grow to global

2.3.2.3 Identify System Drivers

System drivers are elements of design that dominate trade studies. Some are requirements, or needs of customers, such as low price. Some are outside constraints, such as regulatory issues. Other constraints are engineering factors such as temperature extremes in orbit. Most system drivers are obvious early in the design process; however, early identification of system drivers will enhance the effectiveness of the design process. Everyone involved in the design process must pass along perceived design drivers to senior systems engineers as soon as possible to ensure that optimum approaches can be developed and implemented. Risk mitigation efforts must start as soon as possible to affect project success. Preliminary design drivers for three projects are shown in Table 2.13.

Table 2.11. System Drivers

Little LEO	Big LEO	Super GSO
Quality standards	Quality of voice	BER demands
Frequency allocated	Frequency choice	Frequency chosen
Numbers of users	Link Margin	Number of users
Number of satellites	Number of satellites	GEO range to horizon
Range to devices	Altitudes of orbits	Complexity of satellites
Coverage	Number of satellites	Size of Satellite
Processor sizing	Processor speeds	Maximum time delays
Up/down link needs	Crosslinks loading	Link Margin & rain fade
Size of devices	Subscriber unit size	Voice quality
# of distributors	Number of gateways	Ground antenna size
	Number of subscribers	Data rate to users
	Launch availability	Launch vehicle availability
		Operations Center Complexity
		Interconnection to regional PSTNs[2]

One of the principal design drivers is orbital altitude. Therefore, the following chart describes attributes of each orbital range. Breakout is between LEO (lowest, medium, and high), MEO and GSO. Additional historic orbital regimes (elliptical and Molniya) for communications satellites are discussed as they are used for distinguishing between GMSS options.

---

[2] Public Switched Telephone Network

Table 2.12. Altitude Characteristics

	Advantages	Disadvantages
Low Earth Orbit 150-500 Kms Low	Closest to the subscriber units Lowest cost to orbit	Most satellites required for continuous coverage Greatest drag
Low Earth Orbit 500-1,000 Kms Medium	Inexpensive to orbit Good visibility Low radiation and drag	Needs many satellites
Low Earth Orbit 1,000-2,000 Kms High	Excellent coverage Less satellites needed	More expensive to orbit Radiation starting to effect design
Medium Earth Orbit 2,000-10,000 Kms	Minimum # satellites required Longer dwell times	In radiation environment More expensive to orbit
Elliptical Apogee < 10,000 Kms	Focused coverage Excellent dwell times	Unique operational aspects Expensive to orbit
Elliptical with high Apogee	Tremendous coverage of northern tier area Long dwell times (12 hour repeating traces)	Very expensive to orbit Regional per set of satellites Large ground antenna High radiation environment
Geosynchronous Earth Orbit	24 hour dwell time High data rate to large staring antennas Voice data rates to small mobile apertures Least numbers of satellites Simple antenna	Greatest path time delays Very expensive to orbit Regional per set of satellites

2.3.2.4 Mission Architecture

The process of creating and then eliminating multiple architectures and concepts until a single (or principal and backup) concept is selected is iterative and based primarily upon requirements satisfaction. This characterization of a Mission Concept and Single Systems Architecture is usually complex and time consuming. Trade studies at this top level must compare values based upon customers needs as well as engineering expertise and knowledge. One of the first steps within this process is evaluating system drivers against each other and trading results against customers' desires and project completion success. Most systems architects and systems engineers have difficulties during this phase as there is so much uncertainty and so little fact. Trade spaces often have large holes to fill with numbers whose accuracy ranges greater than themselves. Usually, toward the end of this process step, the mission concept is more defined with a system architecture in draft form - frequently with a principle and a few backups. This usually resides in a Systems Concept Document and is presented and finalized at the Preliminary Design Review.

Table 2.13. Principle Design Factors

Little LEO	Big LEO	Super GSO
Orbit chosen	Orbit chosen	Orbit chosen
Number of satellites	Number of satellites	Number of satellites
Size/weight of S/C	Size/weight of S/C	Size/weight of S/C
Processor requirements	Processor requirements	Processor requirements
Subscriber unit size	Subscriber unit size	Subscriber unit size
Schedule and Cost	Network supporting	Network supporting
	PSTN connectivity	PSTN connectivity
	Schedule and Cost	Schedule and Cost

2.3.3   Evaluation and Review

The first step in this process would be to update the requirements' documents based upon the accrued knowledge gained during the first two activities.

2.3.3.1   Review Requirements Specifications

Requirements development was shown as an early step in the process. The architectural process encourages iterative attempts at this phase as understanding the requirements will increase the probability for success. Customers, stakeholders and investors are usually very involved in this step with a major milestone at the Systems Requirements Review.

2.3.3.2 Develop a Vision

These processes of design for complex and sophisticated projects now require that diverse clients, stakeholders, customers, project engineers, manufacturing engineers, systems engineers all see the same program description and concept. A space systems architect through the establishment of a system's vision can facilitate this common view. This should reflect what must be. The vision should be a statement that can engender belief and commitment. As the mega-project concept develops in size, complexity, number of players, and schedule, the vision should provide consistent direction based upon common beliefs and expectations.

Table 2.14. Vision

Little LEO	Big LEO	Super GSO
Global Messaging	Global Personal Communications Anyone, Anywhere, Anytime	Regional mobile voice and data.

2.4 Summary

This chapter was designed to explain the architectural process and show how the three categories of the second generation GMSS fit together. The process you just walked through, dealt with the tremendous uncertainties that face business teams who take on the challenges of GMSS. Some systems are operational. Some systems did not survive. As a result, the industry is still under dynamic pressures to succeed. A comparison of operational, under development, and paper systems illustrates the exciting aspects of Global Mobile Satellite Systems. Further, paper systems are shown as either with contracts and significant funding, or just filings with very little recognizable progress. Each of the three system architectures in this chapter (Orbcomm for Little LEOs, IRIDIUM for Big LEOs, and ACeS for SuperGSOs) are used to illustrate various trades conducted during their development and are representative of their category.

REFERENCES

BOOKS
Johnson, Nicholas L. and D. s. McKnight. 1991. *Artificial Space Debris*. Malabar, Florida: Krieger Publishing Company.

Orbital Debris, A Technical Assessment. 1995. Washington, D.C.: National Academy Press.
Rechtin, Eberhardt and Mark W. Maier. *The Art of Systems Architecting.* 1997. Boca Raton: CRC Press.

Rechtin, Eberhart. *Systems Architecting, Creating and Building Complex Systems.* 1991. Englewood Cliffs, New Jersey: Prentice Hall.

REPORTS
Globalstar Communications Magazine, Special Edition for Telecom 1999 – Geneva, Switzerland, Vol. 8, Oct. 1999.

IRIDIUM© Today. Spring 1996, Vol. 2, # 3.

JOURNAL/SYMPOSIA ARTICLES
Stone, Arthur G. III, "Architecture Systems Engineering Process."

Swan, Peter A. "The IRIDIUM© Story - Where will it Lead? A De-Regulation and Commercialization Success," (IAF-97-M.1.01), 48th Congress of the International Astronautical Federation, Turin, Italy, 6 October 1997.

Swan, Peter A. "Space Systems Architecture: A Constellation Necessity," (IAF-97-M.3.01), 48th Congress of the International Astronautical Federation, Turin, Italy, 6 October 1997.

Swan, Peter A. "Systems Engineering Trades for the IRIDIUM© Constellation," JOURNAL OF SPACECRAFT AND ROCKETS, Vol. 34., No. 5, Sept-Oct 1997. Co-Authors Garrison, Ince, Pizzicaroli.

Maine, Kris; Devieux, Carrie, and Swan, Pete "Overview of IRIDIUM© Satellite Network," IEEE Western Communications Satellite Systems Conference, San Francisco, California, November 1995.

Swan, Peter A. "Manufacturing Technologies, the "Key" to a 66 Small Satellite System," (IAF-94-U.3.475), 45th Congress of the International Astronautical Federation, Jerusalem, Israel, 9 October 1994.

Swan, Peter A. "77 to 66 - The IRIDIUM© Improvement," (IAF-93-M.4.339), 44th Congress of the International Astronautical Federation, Graz, Austria, 16 October 1993.

Chapter 3

Market Demand Considerations

Carrie L. Devieux Jr.[1] and George Besenyei[2]
[1] Chandler, Arizona,   [2] Cave Creek,, Arizona

3.1   INTRODUCTION

Market demand trends for Global Mobile Satellite Services (GMSS) are examined in this chapter. Figure 3.1 depicts a typical, generic, market demand curve, sometimes referred to, as the "S-Curve". The terrestrial cellular business has been on this curve since the early 1980s when the metropolitan test service was deemed successful. When a service/product is first introduced on the market, there is usually a period of latency, i.e. a "period of awareness" by potential customers, where the service is being extensively advertised. Customers, who purchase the service/product at this stage, are usually "early-adopters." Prices for handsets, monthly service charges, and per-minute service charge are usually fairly high and tend to limit customers to businesses and high-income class individuals. Early terrestrial cellular phones were large and bulky and were designed to be vehicle mounted. These early services used the analog modulation access techniques (Frequency Division Multiple access or FDMA). As the benefits of the new service/product become better known, the number of subscribers/customers begins to increase, slowly at first. The potential customers ask themselves whether they need this product, at what price, and what features they like about it (service quality, size of handsets, talk time, idle time before recharge...). As demand increases, prices begin to drop. This phenomenon produces further demand increase (the so-called "Price Elasticity Factor") and a period of rapid growth. The onset of this period is usually marked by the introduction of new technology (digital) resulting in smaller, more efficient and lower cost equipment. Changes in the system infrastructure (cell size, base station...) result in improved service quality (fewer dropped calls...). As more customers come on-board, service charges will drop, further accelerating demand. Eventually, demand will reach a saturation level, indicating that a substantial portion of the addressable market has been captured, and/or that various competitive products are siphoning out demand. This is the replacement period where customers are getting rid of older equipment and replacing them with some having better features. Customers are migrating from one service provider to another (the "churn phenomenon") in the hope of getting better deals. Because of advancing technology, new products are being introduced with better capability/quality, and lower price, which tend to render the original service/product obsolete. Subsequently, market demand starts declining.

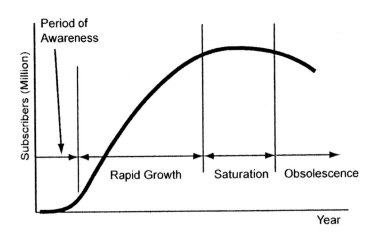

Figure 3.1 Generic Market Demand Curve

3.2   HISTORICAL MARKET DEMAND FOR VARIOUS SERVICES AND CONSUMER PRODUCTS

The original cellular architecture was introduced in 1971 by the AT&T Bell Laboratories, Murray-Hill, N.J.. The system named Advanced Mobile Phone System (AMPS) used the analog modulation technique. It was designed to provide higher capacity than the existing Private Mobile Radio (PMR) system, through the use of frequency reuse via a cell structure (a large coverage area replaced by several interconnected smaller areas or cells)

(Stuber 2001; Oliphant 1999). In summary, the introduction of AMPS was characterized by:

• In the U.S.A., the first generation AMPS launched in 1983, as a metropolitan area test in Chicago. Phone prices and airtime charges were relatively high, resulting a relatively low demand. In the 1990's, digital technology was introduced, leading to rapid demand growth for digital PCS (Personal Communications Systems) services. Many years have passed between the original concept introduction by ATT (1971) and the implementation and launch of services, in part, due to various FCC (U.S. Federal Communications Commission) regulatory issues.

• A period of time where regulatory and technical barriers needed to be smoothed out.

• A period of awareness when the public begins to assess the need for such new services. The early systems (so-called First Generation or 1G) suffered the effects of multipath fading leading to dropped calls or system overloading caused an unacceptable call-access success rates.

• The introduction of digital technology in the early 1990's allowing higher spectrum efficiency, smaller phones, lower prices as well as better voice quality and many practical features appealing to the user. Digital technology (GSM, TDMA, CDMA, PDC) significantly reduced the cost of handset and infrastructure. Figure 3.2 (Lucent Technologies, Qi Bi et al. 2001) shows how the decrease of the number of integrated circuits (ICs) per terminal coupled with the advantages of digital modulation, corresponds to the substantial increase in demand. The use of ASICs (application specific integrated circuits) led to a considerable decrease of handset size. As a result, Personal Communications Systems (PCS) became possible making cellular phones much more appealing than the vehicular communication services (with roof-top antennas).

3.2.1 Cumulative Global Terrestrial Cellular Usage and Market Projections

Figure 3.3 is a summary of cumulative global cellular demand projection as published in 1999 by the Cahners In-Stat Group. The figure shows the rapid growth predicted globally for various terrestrial cellular standards. Since that time, new services have begun to appear for Mobile Internet and WEB services [Japan (NTT-DoCoMo) and others in Europe, U.S.]. The GSM System, which employs TDMA (Time Division Multiple Access) (Mehrotra 1997), is growing rapidly worldwide (Europe, Asia, U.S.).

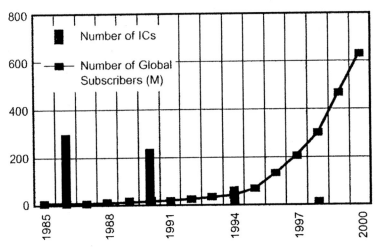

Figure 3.2 Global Subscriber Growth and IC Reduction in Terminals
[Lucent Technologies, Bi Qi, January 2001]

Also gaining in popularity are the CDMA (Code Division Multiple Access) systems (IS-95 and others in U.S, Asia). The Advanced Mobile Phone System, which employs TDMA systems, is mostly popular in the U.S. and Japan (IS-136 Standard in U.S.; others in Japan). The market for analog systems is seen decreasing rapidly as it is being replaced by the digital revolution.

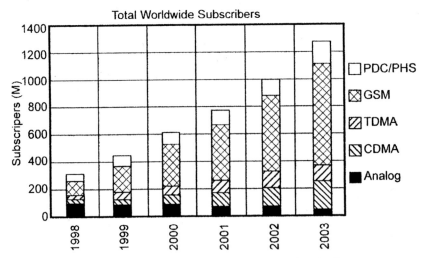

Figure 3.3. Global Cellular Usage and Projections
[Cahners-In-Stat Group-WEB Site 1999]

The first systems to appear were the analog systems called First Generation Cellular, or 1G. The digital systems, the 2G systems, include: PDC/ PHS (Japan); GSM; TDMA; CDMA. The 3G Systems will also be digital, but will handle high-speed data for multimedia applications.

The GSM Association publishes the total number of cellular users each year. The cumulative number is shown in Figure 3.4, together with the yearly increments (columns) for 1992 through 2002 (end-of-year). Figure 3.5 shows the market growth of digital technologies worldwide excluding the GSM users. In December 2000, GSM had 440.7 million users worldwide.

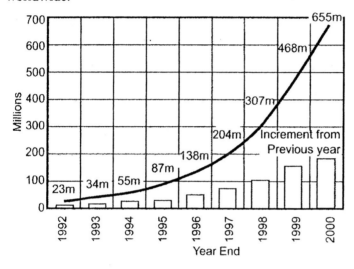

Figure 3.4 Global Cellular Growth [GSM Association / EMC Corp. February 2001]

3.2.2   Growth of U.S. Cellular Since its Inception

This section examines the growth of terrestrial cellular in the U.S. since its introduction in 1983. Figure 3.6 shows the cumulative number of subscribers while Figure 3.7 shows the average monthly bill per user. These data are published by CTIA (The Cellular Telecommunications & Internet Association). At the end of 1993, the monthly bill drops to $61.49 from $68.51 at the end of 1992 (a decrease of 10.2%). The number of subscribers increases to 16,009,461 from 11,032,753 (an increase of 45.1%). Other factors contributing to this growth are the drop of handset prices as well as better service quality and other features such as battery talk time.

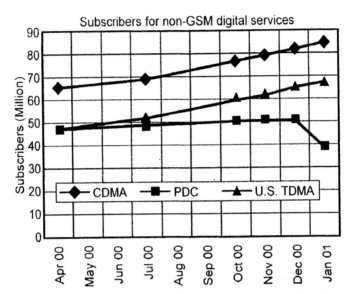

Figure 3.5 Subscriber Growth for "non-GSM" Digital Technologies
[GSM Association –EMC Corp. February 2001]

The Price Elasticity Factor (Samuelson 1995) allows one to assess whether a drop in price will produce a demand increase. Price Elasticity (γ) is obtained from the formula:

$$dN/N = - \gamma * (dP/P)$$

N is the number of subscribers at the end of the previous year; dN is the increase this year. P is the average monthly service charge in the previous year, while dP is the decrease in service charge this year.

3.2.3 Demand for other Consumer Electronics and Services

In this section, the historical behavior of the demand curve is examined for other consumer electronics. We consider the market penetration versus home equipment cost for: In door / outdoor receive units for Satellite Direct Broadcast TV (DBS); Video Cassette Recorder home equipment and the COMPACT Disk (CD) home equipment. On Figure 3.8, the penetration rate (i.e. the fraction of the addressable market acquiring the service or device) is shown for U.S. Direct Broadcast Service (DBS), VCR, and Compact Disk (CD) markets. The year of initiation of service is shown.

U.S. Domestic Subscribership: June 1985-June 2000

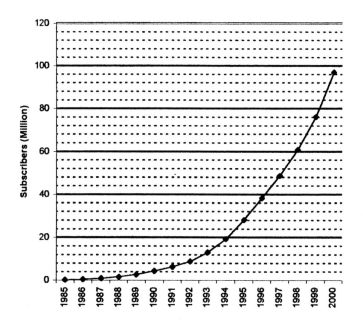

Figure 3.6 Cellular Growth in the U.S.
[CTIA 2000]

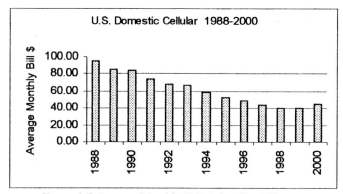

Figure 3.7 Average Monthly U.S. Subscriber Bill
[CTIA-2000]

For example, the VCR was introduced in 1975 and took 8 years to reach a 5% of penetration (1983). Today, the VCR is available in almost every household. More recently DVD recorders have been introduced. They provide much better quality television recording. As the unit price goes down, DVDs have been replacing VCRs which will eventually shift into obsolescence. This process has started in 2001-2002. Figure 3.9 shows the variation of the home equipment prices from year-1 to year-10. The VCR Recorder-Player was quite expensive in the early days (close to $1800), a factor which certainly limited its popularity. In addition, that was compounded by a lack of standards. The curves indicate that unit prices need to drop to below $500 for demand to begin to appreciate. Demand for CDs started climbing around year-4 when their price dropped close to $300 and accelerated around year-10 when prices approached $200. DBS services in the U.S. have shown a moderate but steady growth. In addition to customer equipment cost, an important factor for the DBS market is the monthly service charge. DBS service charges were adequately priced to attract customers. This strategy can only work, of course, if the market is highly elastic: a drop in prices produces a very large increase in the number of subscribers allowing a reasonable profit margin. A similar situation exists for terrestrial cellular and mobile satellite services. In all cases, price elasticity for a given service will depend greatly on the competitive pressure. The new services should be able to provide features and quality of service which will attract customers. For cellular services, factors like size and appearance of the phones, voice and service quality are very important (e.g. minimum number of drop-outs; high probability of establishing a connection in an acceptable time).

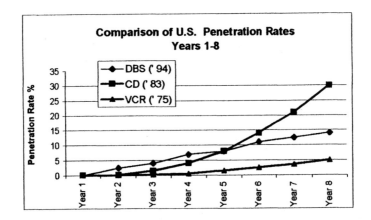

Figure 3.8 Comparative Penetration Rates
[Satellite News 10/2/2000 Phillips Satellite and Space Group]

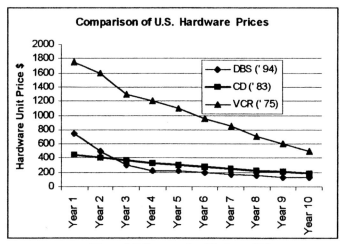

Figure 3.9 Comparative Hardware Prices
[Satellite News 10/2/2000-Phillips Satellite and Space Group]

3.3 GLOBAL MOBILE SATELLITE SYSTEMS

3.3.1 INMARSAT

The INMARSAT System initiated the first mobile satellite service. INMARSAT was organized in London in 1979, with the purpose of providing communication linkage to ships at sea (about 10,000 ships). Members of the original consortium were various countries with interest in commercial shipping. The satellites use the Geostationary orbit and have evolved throughout the years: Comsat (Marisat), ESA (Marecs), INTELSAT MCS, INMARSAT II, INMARSAT III, and INMARSAT IV. Initially, services were only provided over the oceans because of limited antenna coverage. Eventually, satellite antenna coverage of the newer satellites was extended to land areas as well. Transmission capability was also improved to allow the use of smaller shipboard stations. [The original ship equipment was quite bulky]. The newer ship terminals (INMARSAT-A) are substantially reduced in size.

Land communication was made practical by transportable terminals such as the INMARSAT-M which is of the size of a briefcase. The more recent Mini-M is portable and of the size of a Laptop Computer. It weighs about 2.4 Kilograms and it can handle voice and data (2.4 Kilobits/sec). A Mini-M costs about $ 2,500 (Space News: August 1999). Initially retail service charge was as high as $15 per minute, but service was much more reliable than the HF services available at that time. Consequently demand grew steadily for INMARSAT services.

Figure 3.10 shows the yearly revenue of INMARSAT. Also shown are the INMARSAT-M Air Rates ($/minute). INMARSAT–II satellites considerably increased system capacity. The original INMARSAT system cost $478 million (4 GEO Satellites with 250 circuits each). In 1996, the INMARSAT-III system was launched at a cost of about $807 million (5 GEO satellites with 2000 circuits each). At the end of 1998, INMARSAT had more than 140,000 terminals in service (voice & data). More than half were land terminals. A smaller number was used by aircraft; however, 75% of the revenue still came from maritime users. INMARSAT is a success story. Figure 3.11 shows the average usage (minutes) per user per month. INMARSAT has plans for a 3G satellite, INMARSAT-Horizons satellite, at GSO, which will deliver 144 Kilobits/ second to its larger users.

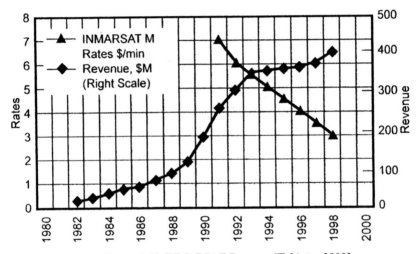

Figure 3.10 INMARSAT Revenue [TelAstra-2000]

3.3.2   Second Generation (2G) Global Mobile Satellite Systems

The First Generation (1G) of Mobile Satellite Systems (MSS) consists of GSO satellites using conventional transparent transponders. Moreover they do not use more advanced technologies such as spot beams. These limitations do not allow hand-held devices similar to Big LEOs and other 2G Systems.

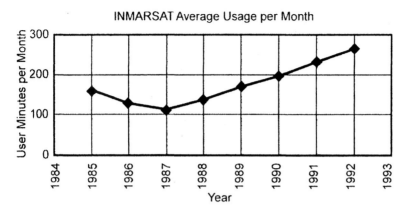

Figure 3.11 INMARSAT Average User Minutes per Month per User

On the other hand, they involve lower cost and more flexibility in adjusting to services and bit-rates to a changing market place. The 1G systems have been providing niche services for a number of years with limited, albeit adequate, capacity for the communities they are serving. They include INMARSAT (primarily maritime services); Mobilesat (Optus) of Australia; AMSC (now Motient Inc.) of the U.S.; TMI of Canada; and others in Europe and Japan. They provide services to ships, boats, trucks, cars and more recently to Land-Mobile Users (INMARSAT). INMARSAT was initially formed to provide highly needed reliable communication to the niche market of commercial maritime shipping. INMARSAT III has narrower regional spot beams in addition to the "global" beams, to allow the use of transportable/portable for Land Mobile Services, such the INMARSAT-M and Mini-M terminals. Some of the INMARSAT terminals were used quite heavily during the Gulf War (1991) for news-gathering functions by the global news media. The terminals used a deployable dish antenna. The type-A terminals use a small deployable dish antenna. The INMARSAT-M (Nera, Norway) terminal can transmit up to 64 Kilobits per second and weighs about 4 Kilograms. It initially cost $11,000 (Space News August 9, 1999). The Mini-M (NEC, Japan) can transmit 2.4 Kilobits per second, similar to the 2G IRIDIUM handset. It weighs 2.4 Kg as compared to original IRIDIUM phone (454 grams) and the Globalstar phone (375 grams). The original Globalstar phone can handle 9.6 Kilobits per second. The Mini-M costs about $2,500 (Year-2000).

Figure 3.12 shows the actual demand growth for GMSS until the end of 1998 (voice users on 1G). INMARSAT had about 80,000 voice users. The TelAstra 2000 forecast estimated projected demand for 1999 to 2005 as shown in the figure. Also shown, is an earlier prediction (1997) by Merrill Lynch. Various market research firms have also generated similar projected demand curves. The sharp growth rate was the projected contribution of various 2G GMSS systems under development. IRIDIUM started operation in November 1998, and had about 67,000 users when it filed for bankruptcy in August 1999. In December 2000, the assets were acquired by IRIDIUM Satellites LLC., which initiated service on the 1st of April 2001. Globalstar started operation in the Spring of 2000. The third Big LEO, ICO of London, filed for bankruptcy in 1999, prior to launching any of its planned constellation. It has since been reorganized as NEW ICO and has modified its business plans to provide multimedia services (3G) capability. Satellites are being up-graded for that purpose.

Figure 3.13 shows the annual revenue of INMARSAT as compared to an earlier IRIDIUM projection. INMARSAT has been in operation since 1982, and had about $400M revenue at the end of 1998. IRIDIUM LLC projected about $450M at the end of 1999. For reference purposes, a Total GMSS Revenue Projection published in 1997 is also shown. Figure 3.14 shows how the average U.S. monthly cellular bill has decreased from about $100 at the end of 1998 to about $40 at the end of 1999. Originally, IRIDIUM expected about $200 per month per user in the 1998-1999 time frame and Globalstar expected about $100 per month per user in the 1999-2000 time frame.

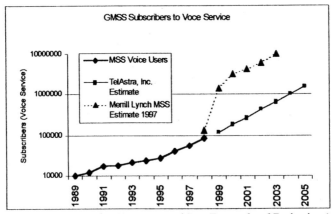

Figure 3.12 Voice Users (Actual Past Demand and Projections)
[TelAstra, Inc. 2000]

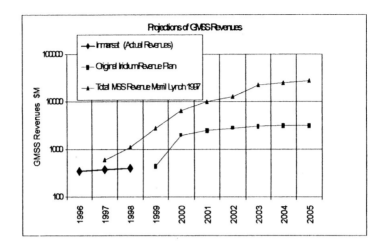

Figure 3.13 Projections of GMSS Revenue [TelAstra 2000]

3.4   COMMENTS

As of the end of 2000 and beginning of 2001, the market for the Big LEO services was still being firmed up. Progress is being hampered by the price of handheld units and the service charge. The trend seems to be in the direction of possible niche markets where sufficient number of subscribers can be found. The Super GSOs (ACeS and Thuraya) seem to be moving in that direction. ACeS, for example, is a regional system covering portions of Asia with a high population density. Their service charges are lower than that of the Big LEOs. Eventually, many of these 2G systems should find the proper niche market to become a financial success. Thuraya has recently introduced a 144 Kbps service in cooperation with INMARSAT, which is a step towards 3G.

Competitive pressure is from the terrestrial cellular services as well as within GMSS. However, satellites have the advantage of ubiquity and could be put to advantage in serving large populations without an adequate telephone infrastructure.

Future trends appear to be in the direction of high speed, multimedia services (144 kilobits/second). INMARSAT is also developing INMARSAT-IV (Horizons), which will be a GSO satellite, and provide 144 Kbps and 432 Kbps, multimedia services. INMARSAT expects to provide service at $1 per minute. Many improvements could help the GMSS gain the edge. However, they may not be all financially realizable.

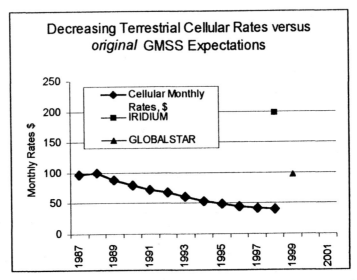

3.14 Decreasing Cellular Rates Compared to MSS Expectations [TelAstra, Inc. 2000]

As the global economy improves, these systems will have a better chance to prosper. Desirable developments include smaller and less expensive handsets and the ability to operate inside a building as well as other improvements.

Caution is always advisable when dealing with major new investment programs. This point is put into evidence by the following GMSS market projections generated at three different points in time: before, during, and after the IRIDIUM bankruptcy proceedings. [1997 Outlook, State of the Space Industry – Space News, May 4, 1998 "Global Satellite Marketplace '98" – 2000 State of the Space Industry]

Table 3.1 Three GMSS Market Projections - Revenues ($ Billion)

	97	98	99	00	01	02	03	04	05
Estimate in '97	.7	1.1	3.4	7.1	9.8	13.3	23.3	27.3	29.2
Estimate in '99		.05	.4	1.5	3.7	6.8	10.3	14.1	17.5
Estimate in '00			.4	.7	1.1	1.5	2.4	3.0	3.8

REFERENCES

BOOKS

Lee William C. "Mobile Cellular Telecommunications," McGraw-Hill, Inc. 1995.
Mehrotra Asha. "GSM Systems Engineering." Artech House Publishers. 1997.

Samuelson Paul A. and Nordhaus William D., "Economics", M.I.T   Textbook, McGraw-Hill. 1995.

Stuber Gordon L. " Principles of Mobile Communication " Second Edition, Kluwer Academic Publishers, 2001.

REPORTS
Rusch Roger R. "Investing in Mobile Satellite Services." Report by TelAstra, Inc. Palos Verdes, CA 90274, February 2000.

"1997 Outlook, State of the Space Industry," Space Publications, Reston, Va, 1997.
"State of the Space Industry, 2000," International Space Business Council, 2000.

JOURNAL/SYMPOSIA ARTICLES
Oliphant Malcolm W. " The mobile phone meets the Internet," IEEE Spectrum, August 1999.

"Global Satellite Marketplace '98," a Study by Merrill Lynch, Space News, May 4, 1998.

Qi Bi, Zysman George I., Menkes Hank, "Wireless Mobile Communications at the Start of the 21st Century", IEEE Communications Society Magazine, January 2001.

Chapter 4

Regulatory and Spectrum Considerations

Donald Jansky[1] and Robert D. Kubik[2]
[1]*Jansky / Barmat Telcom, Inc.,* [2]*Washington, D.C.*

4.1 SATELLITE FREQUENCY ALLOCATIONS

Mobile satellite communications system frequency allocations, and regulations, fall under the influence of a combination of three types of entities; an administration, a group of administrations or an international body. Each administration (or nation) has a sovereign right to specify the conditions of operation that a satellite system must follow in order to provide service within the border of that administration. Since the region of service of a satellite system typically encompasses many administrations, often they form groups to work together in order to develop rules/regulations for these systems to ensure a common regulatory regime. One example of this type of group is the European Conference of Postal and Telecommunications Administrations (CEPT) that is developing harmonized allocations throughout Europe. The final group is an international body that considers all regions of the world, the International Telecommunication Union (ITU) and gathers all administrations to develop a common set of allocations.

4.1.1 International

The ITU is affiliated with the United Nations (UN) organization. It implements the treaty that contains the international allocation of frequencies. These allocations are incorporated in the Radio Regulations. Changes to these international allocations are developed and modified at

World Radio Conferences (WRC) that meet every 2-3 years, which are treaty conferences. As a result of a WRC, changes are proposed to the Radio Regulations, and once ratified by each administration, these changes are treated like any other treaty. The ITU table of allocations, which are part of these regulations, specifies the bands of frequencies for different services in three distinct Regions of the earth as shown in *Figure 4.1* (ITU RR S5.2).

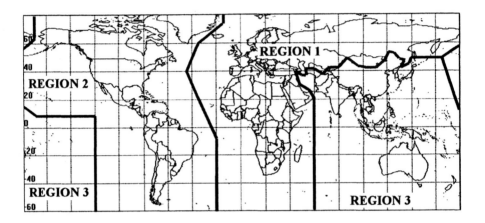

Figure 4.1. ITU Regions (ITU RR S5.2)

Various frequency bands are allocated for the Mobile-Satellite Service (MSS) in the span ranging from 137 MHz to 265 GHz. Other allocations that may be utilized by a MSS system include Inter-satellite allocations and Feeder link allocations. The use of a particular frequency is determined by its status; MSS allocations have either a primary or secondary allocation. A primary service allocation indicates the type of applications that are principally intended to operate in a particular frequency band. Secondary service allocations indicate the type of applications that are also permitted in some bands, but with lower status than a primary allocation. The regulations state that stations of a secondary service 1) shall not cause harmful interference to stations of a primary service and 2) cannot claim protection from harmful interference from stations of a primary service.

Definitions of terms associated with the Mobile-Satellite service are as follows (ITU-R Radio Regulations):

> *mobile-satellite service (MSS): A radiocommunication service*:
> - between *mobile earth stations* and one or more *space stations* or between *space stations* used by this service, or
> - between *mobile earth stations* by means of one or more *space stations*.

This service may also include *feeder links* necessary for operation.

radiocommunication service: A service involving the transmission, *emission* and/or reception of *radio waves* for specific *telecommunications* purposes.

mobile earth station: A *earth station* in the *mobile-satellite service* intended to be used while in motion or during halts at specified points.

land mobile-satellite service: A *mobile-satellite service* in which *mobile earth stations* are located on land.

maritime mobile-satellite service: a *mobile-satellite service* in which *mobile earth stations* are located on board ships; *survival craft stations* and *emergency position-indicating radiobeacon stations* may also participate in this service.

aeronautical mobile-satellite service: A *mobile-satellite service* in which *mobile earth stations* are located on board aircraft; *survival craft stations* and *emergency position-indicating radiobeacon stations* may also participate in this service.

earth station: A *station* located either on the Earth's surface or within the major portion of the Earth's atmosphere and intended for communication:
- with one or more *space stations*; or
- with one or more *stations* of the same kind by means of one or more reflecting *satellites* or other objects in space.

space station: A *station* located on an object which is beyond, is intended to go beyond, or has been beyond, the major portion of the Earth's atmosphere.

station: One or more transmitters or receivers or a combination of transmitters and receivers, including accessory equipment, necessary at one location for carrying on a *radiocommunication service*, or the *radio astronomy service*.

feeder links: A radio link from an *earth station* at a given location to a *space station*, or vice versa, conveying information for a specific *space radiocommunication service* other than for the *fixed-satellite service*. The given location may be at a specified fixed point, or at any fixed point within specified areas.

4.1.2   Regional

Regional allocations have developed due to the requirement of a region, for example Europe, to provide service. For example in Europe a GSO   (Geostationary Orbit) system could fully utilize all of its capacity to provide Mobile Satellite service in that region. In these cases the regional allocations may differ from those found in the ITU Radio

Regulations. This example is currently being developed by CEPT via harmonization frequency allocations. CEPT has endorsed the principle of adopting a harmonized European Table of Frequency Allocations and Utilization by the year 2008 (ERC Report 25). This activity is being performed by the CEPT ERC's European Radiocommunications Office (ERO) through a series of Detailed Spectrum Investigations (DSIs) which consider the harmonization of different frequency ranges. The frequency range 960 MHz to 3400 MHz has not yet been covered by a DSI. However, the frequency range 860 MHz to 3400 MHz covered by DSI Phase III takes advantage of the work already being undertaken by CEPT for the frequency band 1350 - 2690 MHz by extending the frequency range to cover this broader band. The frequency range 29.7 MHz to 960 MHz was covered by DSI Phase II and the range 3.4 GHz to 105 GHz was covered by DSI Phase I which has lead to an agreed frequency allocation and utilization plan.

4.1.3 Domestic

Domestic allocations to the Mobile Satellite Service are typically guided by the regional and/or international allocations. Differences occur due to the time lag between when a World Radio Conference completes it work and the amount of time that it takes an administration to implement changes, if warranted. Often these changes are not warranted due to domestic allocations (or services) in which the administration does not want a competing Mobile Satellite System to possibly share frequencies. In such cases the individual administration would not implement allocation changes.

4.1.4 Proposed Utilization of Frequency Allocations

Many Mobile Satellite Systems are currently operating, or plan to operate in the future, that use many of the aforementioned frequency allocations. The utilization of frequencies of the service link (communication to/from the mobile earth station) for some of the systems under consideration is shown in *Table 4.1*.

Table 4.1. Utilization of Mobile Satellite Service Allocations (ITU-R Recommendations, M series)

System	Service Link (Earth-to-space / space-to-Earth)	Type
Little Leo		
Orbcomm (Leotelcom-1)	148 – 150 MHz / 137 – 138 MHz	Global
	148 – 148.855 MHz / 137.0725 – 137.9275 MHz	Global
E-SAT (Leotelcom-2)	148 – 150.05 MHz / 137 – 138 MHz	Global
Leo one USA		
Other		
MLMS	387.25 – 388.75 MHz / 400.15 – 401 MHz	Global
SAFIR 2	399.9 – 400.05 MHz / 400.6 – 400.9 MHz	Global
Big Leo		
Globalstar	1610 – 1621.35 MHz / 2483.5 – 2500 MHz	Global
Ellipsat	1610 – 1621.35 MHz / 2483.5 – 2500 MHz	Global
Odyssey	1610 – 1621.35 MHz / 2483.5 – 2500 MHz	Global
Constellation	1610 – 1621.35 MHz / 2483.5 – 2500 MHz	Global
AMSC	1610 – 1621.35 MHz / 2483.5 – 2500 MHz	Reg.
Iridium	1621.35 – 1626.5 MHz / 1621.35 – 1626.5 MHz	Global
Super GSO		
ACeS	1626.5 – 1660 MHz / 1525 – 1559 MHz	Reg.
Thuraya	1626.5 – 1660 MHz / 1525 – 1559 MHz	Reg.
2 GHz		
ICO	1990 – 2020 MHz / 2165 – 2200 MHz	Global
Macrocell (Iridium LLC)	1980 – 2025 MHz / 2160 – 2200 MHz	Global
	1980 – 2025 MHz / 2160 – 2200 MHz	Global
GS-2 (Globalstar)	1990 – 2025 MHz / 2165 – 2200 MHz	Global
Ellipso 2G (MCHI)	1980 – 2025 MHz / 2165 – 2200 MHz	Global
Constellation II	1970 – 1990 MHz / 2160 – 2180 MHz	Reg.
PCSAT (AMSC)	1990 – 2025 MHz / 2165 – 2200 MHz	Reg.
CellSat	1990 – 1998.25 MHz / 2165 – 2173.85 MHz	Global
The Boeing Company	1990 – 2025 MHz / 2160 – 2200 MHz	Reg.
TMI Communications	1980 – 2025 MHz / 2160 – 2200 MHz	Reg.
Horizons (Inmarsat)		

4.2   REGULATORY CONSIDERATIONS

Regulatory considerations that are considered in this section are categorized by two approaches that allow sharing between systems in the Mobile-Satellite service and sharing between a system in the Mobile-Satellite service and other services. The first approach is through the development of technical standards that permit sharing; aspects include but not limited to satellite downlink power flux density limits (or thresholds) and out-of-band limits (or thresholds). In the cases in which the technical standard is a threshold and cannot be achieved, coordination procedures are invoked that allow the two interested parties (in the case of the ITU the administrations involved) to technically evaluate the sharing aspects of all concerned systems to arrive at mutually acceptable means of sharing.

4.2.1   Technical Standards

Technical standards that are imposed on the Mobile Satellite Service that are found in the ITU are of two types. Those found in the Radio Regulations and those that are in ITU-R Recommendations (ITU-R Recommendations, M series).

Table 4.2 Relevant Mobile Satellite Service Recommendations for systems found in Table 4.1

Recommendation	Description
M.817	International Mobile Telecommunications-2000 (IMT-2000). Network architectures
M.819	International Mobile Telecommunications-2000 (IMT-2000) for developing countries
M.830	Operational procedures for mobile-satellite networks or systems in the bands 1 530-1 544 MHz and 1 626.5-1 645.5 MHz which are used for distress and safety purposes as specified for GMDSS
M.1034-1	Requirements for the radio interface(s) for International Mobile Telecommunications-2000 (IMT-2000)
M.1037	Bit error performance objectives for aeronautical mobile-satellite (R) service (AMS(R)S) radio link
M.1079	Speech and voiceband data performance requirements for International Mobile Telecommunications-2000 (IMT-2000)
M.1183	Permissible levels of interference in a digital channel of a geostationary network in mobile-satellite service in 1-3 GHz caused by other networks of this service and fixed-satellite service
M.1184	Technical characteristics of mobile satellite systems in the 1-3 GHz range for use in developing criteria for sharing between the mobile-satellite service (MSS) and other services using common frequencies
M.1186	Technical considerations for the coordination between mobile satellite service (MSS) networks utilizing code division multiple access (CDMA) and other spread spectrum techniques in the 1-3 GHz band
M.1230	Performance objectives for space-to-Earth links operating in the mobile-satellite service with non-geostationary satellites in the 137-138 MHz band
M.1231	Interference criteria for space-to-Earth links operating in the mobile-satellite service with non-geostationary satellites in the 137-138 MHz band
M.1232	Sharing criteria for space-to-Earth links operating in the mobile-satellite service with non-geostationary satellites in the 137-138 MHz band
M.1389	Methods for achieving coordinated use of spectrum by multiple non-geostationary mobile-satellite service systems below 1 GHz
M.1343	Essential technical requirements of mobile Earth stations for global non-geostationary mobile-satellite service systems in the band 1-3 GHz

Table 4.3. Technical standards in Radio Regulations for the Mobile Satellite Service[ITU-R]

Source	Requirement
S5.364	Mobile earth stations shall not produce a peak e.i.r.p density in excess of −15 dB(W/4kHz) in the part of the band used by systems in accordance with the provisions of S.5366. In the part of the band where such systems are not operating, the mean e.i.r.p. density of a mobile earth station shall not exceed −3 dB(W/4kHz). S5.366: The band 1610-1626.5 MHz is reserved on a world-wide basis for the use and development of airborne electronic aids to air navigation and any directly associated ground based or satellite-borne facilities.
S5.407	In the band 2500-2520 MHz, the power flux-density at the surface of the Earth from space stations operating in the mobile-satellite service (space-to-Earth) shall not exceed −152 dB(W/m^2/4kHz) in Argentina, unless otherwise agreed by the administrations concerned.
S21.8	The e.i.r.p. transmitted in any direction towards the horizon by an earth station shall not exceed the following limits except as in provided in No. S21.10 or S21.11: a) in frequency bands 1-15 GHz +40 dBW in any 4 kHz band for $\theta \leq 0°$ +40 + 3 θ dBW in any 4 kHz band for $0 \leq \theta \leq 5°$; and b) in frequency bands above 15 GHz +64 dBW in any 4 kHz band for $\theta \leq 0°$ +64 + 3 θ dBW in any 4 kHz band for $0 \leq \theta \leq 5°$, where θ is the angle of elevation of the horizon viewed from the centre of radiation of the antenna of the earth station and measured in degrees as positive above the horizontal plane and negative below it.
S21.16	The power flux density at the Earth's surface produced by emissions from a space station, including emissions from reflecting satellite, for all conditions and for all methods of modulation, shall not exceed the limit given in Table S21-4. The limit relates to the power flux density which would be obtained under free space propagation conditions and applies to emissions by a space station of the service indicated where the frequency bands are shared with equal rights with the fixed or mobile service, unless otherwise stated.

Table S21-2 (Portions that list mobile-satellite as the applicable service)

Frequency Band (GHz)	Limit in dB(W/m^2) for angle of arrival (δ) above the horizontal plane			Reference bandwidth
	0° - 5°	5° - 25°	25° - 90°	
3.4 - 4.2 4.5 – 4.8 5.67 – 5.725 7.25 – 7.85	−152	−152 + 0.5*(δ-5)	−142	4 kHz
31.0 – 31.3 34.7 – 35.2 37.0- 40.5	−115	−115 + 0.5(δ-5)	−105	1 MHz

The technical standards that are found in recommendations (see Table *4.2*) are used as guidance in technical studies involving sharing, those found in the Radio Regulations in general must be followed (see *Table 4.3*).

4.2.2 Coordination Procedures

Mobile satellite network coordination activities can fundamentally be divided into two parts, (a) coordination among GSO MSS systems, and (b) coordination between non-GSO MSS systems. To date there has been no coordinations between co-frequency GSO MSS systems and non-GSO MSS systems. The provisions of the ITU-R Radio Regulations in Article S9, edition 1998, govern all of these coordinations. This article combines previous regulations known as Article 11, Article 13, and Resolution 46, as well as regulations affecting the coordination of space telecommunication services (ITU-R Radio Regulations).

4.2.2.1   Coordination of GSO MSS Networks

As the oldest mobile satellite networks, dating from the 1970s, these networks have traditionally been guided by the provisions of Article 11, now Article S.9.17 and associated appropriate Appendix S5 provisions of the Radio Regulations. These regulations provide that before notification of the network can take place, coordination with other co-frequency satellite networks as provided by these procedures must take place. The technical conditions, which trigger coordination in this case, are set forth in Table S5-1 of the Radio Regulations Appendix S5 (see *Table 4.4* below).

Unlike fixed satellite networks, that use frequencies in the microwave part of the spectrum or above and employ high gain antennas, MSS networks cannot achieve 2° spacing in the GSO. The VHF and UHF frequencies used by mobile satellite networks and the beamwidths of the user antennas make co-frequency orbital spacing for MSS systems in the order of 30° or more. Given the scarcity of MSS spectrum, international coordination of GSO MSS satellite networks is extremely difficult.   The approaches to coordination of MSS systems in ITU Region 2 vis-à-vis Regions 1 & 3 have evolved somewhat differently.

Inmarsat, as the leader of GSO MSS spectrum use, is the largest claimant to the available MSS spectrum allocations at L-band frequencies. In Region 2 other GSO MSS systems (e.g., those of the U.S. Canada and Mexico) have had to conduct difficult coordination over many years to achieve some degree of successful co-existence. Such coordination was not achieved until all of the parties, administrations, and organizations involved engaged in multilateral coordination in which certain principles were agreed upon regarding use of available spectrum.

Table 4.4. Technical conditions for coordination [ITU-R Radio Regulations, Article S9.7]

Case	Frequency bands (and Region) of service for which coordination is sought	Threshold / Condition	Calculation Method
A station in a satellite network using the geostationary satellite orbit (GSO), in any space radio-communication service, in a frequency band and in a Region where this service is not subject to a Plan, in respect of any other satellite network using this orbit, in any space radio-communication service in a frequency band and in a Region where this service is not subject to a Plan, with the exception of the coordination between earth stations operating in the opposite direction of transmission.	Any frequency band allocated to a space service, where this service is not subject to a Plan	Value of ΔT/T exceeds 6%	Appendix S8

To overcome this problem, of multiple lengthy coordinations in Regions 1 & 3, the GSO MSS operators, proponents, and administrations agreed on a Memorandum of Understanding (MOU) which provided a basis for access to spectrum. The MOU group meets once a year to review status of access based on these principles. The basic principles are set forth in *Table 4.5.* While respecting the Radio Regulations, these principles go further and relate access to spectrum based on expectation of near term future use.

Table 4.5. Principals for Region 1 & 3 GSO MSS operators

The parties will utilize the GSO MSS spectrum assignments in the frequency bands 1525-1554/1545-1559 MHz and 1626.5-1645.5/1646.5-1660.5 MHz in the most efficient manner practicable.
Prior developed agreements will be reviewed annually, in accordance with the relevant provisions of the ITU Constitution, Convention and the Radio regulations.
Agreements will take into account the plans for deployment of satellites for both existing operational satellites and the planned networks of GSO MSS operators and consider their spectral efficiency.
The necessary consultation will be made with parties in Region 2 who have entered into frequency coordination in order to meet the requirements for the seamless operation of GSO MSS networks in more than one Region, or to ensure capability with networks in Region 2.
In the deployment phase of all new networks, the Parties will accommodate testing of new satellites.

4.2.2.2   Non-GSO MSS Coordination

Coordination among Non-GSO MSS systems also falls into two categories: (a) non-GSO MSS systems below 1 GHz; and (b) non-GSO MSS systems in the range 1-3 GHz. In general, procedures for non-GSO MSS network coordination were provided for by WARC-92 in a new regulation known until recently as Resolution 46. The regulation is now known as S9.11A. Table S5-1A from the RR indicates the bands and services to which this regulation applies. The table also indicates the regulation number, and other services to which the regulation applies. Unlike the GSO MSS, coordination is determined by simple frequency overlap, i.e., if the frequency(s) of one non-GSO MSS network over lap those of another such network or that of another co-frequency service, coordination is required.

To date, the only coordinations which have taken place have involved systems below 1 GHz, and those in the L-band part of the 1-3 GHz MSS allocations. In both cases, the principle method for achieving successful coordination has been band or frequency separation or other words the avoidance of frequency overlap. This approach has been more pronounced at the L-band frequencies than at VHF frequencies. The U.S. Europe, and the Russian Federation have adopted band plans to provide for spectrum for "Big Leo" MSS non-GSO systems. This plan for the U.S. is in *Figure 4.2*. It provides for CDMA systems to share part of the spectrum and for TDMA systems to use the remainder. In principle the CDMA systems are supposed to share the spectrum, but to date the technical basis for accomplishing this have not been established.

In summary, it is likely that future MSS coordinations will be based on some form of band segmentation coupled to a due diligence mechanism which gives preference to MSS systems in operation or expected to be implemented.

4.3   MOBILE SATELLITE SERVICE REGULATION

This section provides information on various topics related to the regulation of mobile satellite systems. These include spectrum scarcity/requirements, coordination activities under the ITU-R, GMPCS, IMT-2000, and World Radio Conferences. . All of these topics are interrelated as a consequence of mobile satellite systems need for access to radio spectrum and in particular the unique use of such spectrum by mobile users.

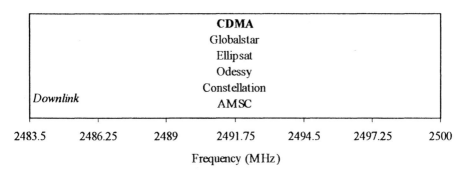

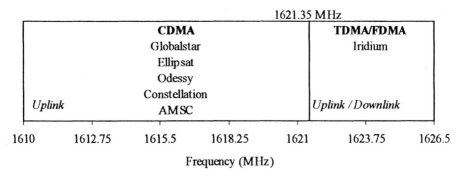

Figure 4.2. US "Big LEO" spectrum allocations.

4.3.1   Spectrum Requirement/Scarcity

The first spectrum allocations for the mobile satellite services were established at the 1971 World Administrative Radio Conference for Space Telecommunications. Additional allocations were made at the 1979 World Administrative Radio Conference. These allocations were sufficient to support the relatively few mobile satellite systems coming into use including Inmarsat and U.S. and Russian government systems. All of these systems were geostationary and in the GMPCS first generation category.

The 1990s have seen this situation change dramatically. The new demand for spectrum to support mobile satellite systems came from non-voice non-GSO, voice non-GSO, and super GSO mobile satellite systems. Some MSS spectrum allocations for these systems were agreed at WARC-92. The impetus for these allocations derives from proposals to provide MSS to hand held terminals. The best frequencies for the types of systems envisioned correspond to frequencies where the technology for the user terminals has been developed. The new MSS allocations were at L-band, S-band, and VHF. The requirement for this spectrum was defined by the multiple proposals put forth for use of the spectrum as a consequence of the "open skies" regulatory policy of the U.S. Federal Communications Commission (FCC). Under this policy when there is a proposal to the FCC for a new type commercial satellite system, the FCC puts this system out for public comment, and inquires as to whether there are other organizations interested in providing this service. This policy fuelled demand for spectrum for MSS systems. At VHF, upon the proposal of the Orbcomm system of Orbital Sciences Corp., there resulted three "little Leo" system proposals including Starsys and Vita. At L-band upon the proposal of Ellipsat, there resulted proposals for four additional non-GSO "Big Leo" Systems, including Iridium, Globalstar, Odyssey, and Constellation. In both cases the combined spectrum requested exceeded the MSS allocations available leading to a situation of "Mutual Exclusivity". In other words, the "open skies" policies created a situation of spectrum scarcity.

The regulatory solution to this dilemma in the United States was to engage in Negotiated Rule Makings. These activities, including government regulators and both satellite and non-satellite communications system proponents, resulted in contentious negotiations, which delayed the introduction of these systems. Nonetheless, solutions were reached on how to share the available MSS allocations among U.S. applicants. Such arrangements did not include the systems of other countries. The scarcity of MSS allocations became apparent in advance of WRC-95, WRC-97 and now WRC-2000. All of the useable MSS allocations at both VHF and L-band frequencies is spoken for by non-GSO or GSO MSS systems. Additionally complications arose because of the late blooming nature of the MSS, the allocations identified for this service at WARC-92 were already being used by other services. VHF compatibility with already existing services was addressed through unique spectrum sharing arrangements such as dynamic channel assignment (DCASS) sharing. L-band sharing was accomplished through band segmentation of modulation types and multiple systems using CDMA modulation. However, the allocations identified at S-band frequencies for MSS have proved to be relatively unavailable because of their shared use with already existing fixed services.

Due to the popularity of the "1st" Round of VHF and L-band MSS type systems, the FCC, in both cases, opened a second round of applicants. In the case of the Big-Leo 2nd round, the allocations to be used were to be those MSS allocations at S-band. In the case of the 2nd round VHF MSS systems, the Commission hoped to obtain additional spectrum at WRC-95 or WRC-97 or share the allocations assigned in the 1st round. Meanwhile, the next generation of GSO MSS systems was developing. These systems known as "Super MSS" would deliver sufficient power to service hand held terminals from the GSO. All of the proposed systems for MSS during the 1990's have given rise to a requirement for MSS allocations in excess of that available in the allocation table today. The requirement is both regional and global and MSS spectrum is scarce for both applications. The global spectrum availability and allocation is especially critical for NGSO systems. Below are requirement statements excerpted from U.S. proposals and ITU documents which explain the nature of the spectrum demand for MSS allocations as it is found. [CPM & US to CPM-99] The difficulty in meeting this demand will continue as a consequence of already entrenched terrestrial services. The lack of availability of MSS allocations has been a contributing factor to the relative non-availability of these systems.

4.3.1.1 Spectrum Requirements for MSS (CPM-99)

4.3.1.1.1 Part A – MSS spectrum requirements in the 1 to 3 GHz range

The need for additional spectrum for the MSS in the 1 to 3 GHz range has been stated in the Conference Preparatory Meeting (CPM) Reports to the last two WRCs. CPM to WRC-95 indicated that a total allocation of between 2 x 75 MHz and 2 x 150 MHz would be required by 2005. Two years later, the CPM to WRC-97 indicated that 2 x 250 MHz would be necessary by 2010.

Two recent submissions to the ITU-R, one a compendium of studies, based on conservative assumptions of MSS traffic growth and realistic MSS spectrum efficiency, have forecast a minimum MSS spectrum demand of between 2 x 125 MHz and 2 x 145 MHz by 2010, depending on the geographic region (CPM Report). [These studies are consistent with the spectrum demand estimates presented in Report ITU-R [IMT.SPEC], which are detailed in the context of the satellite component of IMT-2000. It is important to note that although it is foreseen that some or all of the spectrum requirements for the satellite component of IMT-2000 may be accommodated in the existing MSS allocations, additional MSS allocations are needed to meet the projected total MSS spectrum requirements, since the satellite component of IMT-2000 is only a subset of the mobile service. Indeed, identification of the existing MSS allocations for the satellite component of IMT-2000 places additional emphasis on the need for additional allocations to support both IMT-2000 and non-IMT-2000 MSS requirements. Since all studies indicate that there is a requirement for additional MSS spectrum in 1 – 3 GHz to meet the demand of MSS in time, including the satellite component of IMT-2000, all existing worldwide MSS allocations should be retained for MSS use. Currently the Radio Regulations allocate about 2 x 115 MHz to the MSS in 1 – 3 GHz, with some variations among the ITU Regions. It should however be noted that most of these bands are used by other services in most countries, significantly reducing the actual availability of those bands for the MSS, and in many countries they are not available for MSS at all. This applied in particular to the bands indicated in *Table 4.6*, which are heavily used mainly by terrestrial fixed services in many countries. In order to satisfy the MSS spectrum requirements it is essential to facilitate the MSS access to spectrum already allocated to the MSS, taking into account other services with a co-primary allocation.

Table 4.6. Bands not available for MSS

Band (MHz)	Note
1980 – 2010	
2010 – 2025	Region 2 only
2160 – 2170	Region 2 only
2170 – 2200	
2500 – 2535	
2690	

4.3.1.1.2   Part B – MSS spectrum requirements below 1 GHz

A total of 1.525 MHz (space-to-Earth) and 1.9 MHz (Earth-to-space) are presently allocated on a worldwide primary basis to the MSS below 1 GHz and 300 kHz (Earth-to-space) is allocated for land MSS on a worldwide primary basis. An additional 151.5 MHz may be used subject to agreement obtained under provision RR No.S9.21. It is difficult to implement MSS systems in this 151.5 MHz of spectrum subject to RR No. S9.21 because of the need to obtain agreements with many countries. In addition, 2 MHz (Earth-to-space) in Region 2 is allocated to the MSS below 1 GHz. Some individual countries have additional allocations (Earth-to-space) for the MSS below 1 GHz, appearing in footnotes. These allocations are for both the MSS service links and feeder links.   The Radiocommunication Bureau has identified 22 non-GSO MSS networks as of 28 April 1999, at frequencies below 1 GHz, at some state of coordination under RR No. S9.11A/Resoltuion 46 (Rev.WRC-97). However, it appears that many of the proposed networks cannot be implemented in the existing allocations because there is not enough spectrum readily available without applying S9.21 to allow the development of all of these systems.

It had previously been identified in the CPM Report for WRC-95 and is also stated in *considering* b) of Resolution 214 (Rev. WRC-97) that 7 to 10 MHz of additional spectrum is required for MSS below 1 GHz. An extensive study carried out by an administration in 1996 identified a need for spectrum beyond the current allocations to provide for certain applications until the year 2002. This study identified a spectrum requirement for service links of about 17 MHz on a shared basis, and an additional 4 MHz of shared spectrum for feeder links. In arriving at the conclusions regarding spectrum requirements and market demand, the study took into account the ability of present and future competing terrestrial and satellite technologies to provide these applications. This study estimated that when the data rates and frequency of use among the various users are taken into account on a worldwide basis an average of 3.2 million non-GSO MSS users would be provided service in each 1 MHz of bandwidth for uplinks, and 6.1 million users per MHz for downlinks.

4.3.2   World Radio Conferences (WRCs)

World Radio Conferences result in modifications to the international Radio Regulations (RR). These regulations are considered a treaty which is binding on the member nations of the International Telecommunications Union (ITU). There over 180 member countries. These treaties are applicable to virtually every country in the world. For many countries these constitute their domestic regulations. WRCs are held approximately every two to three years. The modifications to the Radio Regulations, which are considered at these conferences, are normally in accordance with a previously agreed agenda. Considerations on the Mobile Satellite Service have been on the agendas of WARC-71, WARC-79, WARC-92, WRC-95, WRC-97 and WRC-2000. The agenda items that concern the Mobile Satellite Service at this conference are described in *Table 4.7* below (CPM Report).

Table 4.7. Mobile-Satellite Issues at WRC-00

WRC-00 Agenda Item	Topic
1.6	Issues related to the IMT-2000;
1.6.1	Review of spectrum and regulatory issues for advanced mobile applications in the context of IMT-2000, noting that there is an urgent need to provide more spectrum for the terrestrial component of such applications, and priority should be given to terrestrial mobile spectrum needs, and adjustments to the Table of Frequency Allocations a necessary;
1.6.2	Identification of a global radio control channel to facilitate multimode terminal operation and worldwide roaming of IMT-2000;
1.9	Take into account the results of ITU-R studies in evaluating the feasibility of an allocation in the space-to-Earth direction to the mobile-satellite service in a portion of the 1 559 – 1 567 MHz frequency range, in response to Resolutions 213 and 220 (WRC-97);
1.10	To consider results of ITU-R studies in accordance with Resolution 218 (WRC-97) and take appropriate action on this subject; [title of Res. 218]*
1.11	To consider constraints on existing allocations and to consider additional

WRC-00 Agenda Item	Topic
	allocations on a worldwide basis for the non-GSO/MSS below 1 GHz, taking into account the results of ITU-R studies conducted in response to Resolutions 214 (Rev.WRC-97) an 219 (WRC-97);
1.12	To consider the progress of studies on sharing between feederlinks of non-geostationary-satellite networks in the mobile-satellite service and geostationary-satellite networks in the fixed-satellite service in the bands 19.3-19.7 GHz and 29.1-29.5 GHz taking into account Resolution 121 (Rev.WRC-97);

Agenda item 7.2 for WRC-2000 will be to recommend agenda items for the next conference. Some of these agenda items will include issues concerned with Mobile Satellite Spectrum. Each of the agenda items of a conference is prepared for by an ITU Radio Communications Sector Study Group. The ITU-R Sector Study Group 8 is concerned with all Mobile Services. Under this Study Group it is Working Party 8D (WP-8D) which is concerned with issues involving the Mobile Satellite Services (MSS). With input to each of the agenda items, dealing with MSS matters, the WP-8D prepares a technical assessment with respect to the feasibility of the sharing situations being considered with respect to the use of an allocation by the MSS and other services in the same frequency band. For example, the conclusions of WP-8D in relation to the agenda items cited above are indicated in Sections 4.3.2.1 through 4.3.2.4.

4.3.2.1   Agenda 1.6.1

WRC-00 Agenda Item 1.6.1 calls for the review of spectrum and regulatory issues for advanced mobile applications in the context of IMT-2000, noting that there is an urgent need to provide more spectrum for the terrestrial component of such applications, and priority should be given to terrestrial mobile spectrum needs, and adjustments to the Table of Frequency Allocations.

4.3.2.1.1 Frequency Bands 1525 – 1559 / 1626.5 – 1660.5 MHz

These bands are allocated to the mobile satellite service on a global basis. 1525 – 1535 MHz is also allocated on a co-primary basis to the space operation service (s-E). 1525 – 1530 MHz is also allocated in Regions 1 and 3 to the fixed service on a co-primary basis. 1660 – 1660.5 MHz is allocated on a worldwide basis to the radio astronomy service on a co-primary basis. In the bands 1525 – 1559/1626.5 – 1660.5 MHz, MSS systems have been in operation for more than ten years. With an increasing number of new operators and increasing demand for MSS services, the bands are now rapidly approaching saturation, as is evident from the continuing regional multilateral coordination activities. Safety communications have priority in the bands 1530 – 1544/1626.5 – 1645.5 MHz as per RR No. S5.353A, and in the 1545 – 1555/1646.5 – 1656.5 MHz band, as per RR No. S5.357A. The bands 1544 – 1545/1645.5 – 1646.5 MHz are limited to distress and safety communications according to RR Article S31.

Advantages
Although the systems currently operating in these bands were state-of-the-art when launched, technological advancements will continue to lead to new systems being even more spectrum efficient. Hence, as the current MSS systems are eventually replaced by new systems, potentially including satellite IMT-2000 systems, such systems may be able to use these bands. These bands are allocated to the MSS on a global basis.

Disadvantages
The satellite component of IMT-2000 could only be implemented in these bands in the shorter term, taking into account spectrum congestion. Portions of the band 1525 – 1559 MHz and 1626.5 – 1660.5 MHz are subject to constraints. Deployment of IMT-2000 in these bands must take into account that usage in the sub-band 1544 – 1545 / 1645.5 – 1646.5 MHz is limited to distress and safety communications.

4.3.2.1.2   Frequency bands 1610 – 1626.5 / 2483.5 2500 MHz

These bands are currently allocated on a global basis to the mobile-satellite service. The band 1610 – 1626.5 MHz is also allocated worldwide to the aeronautical radionavigation service on a co-primary basis. The band 1610 – 1626.5 MHz is in Region 2 also allocated on a co-primary basis to the radiodetermination satellite service. The sub-band 1610.6 – 1613.8 MHz is also allocated in all 3 Regions to the radio astronomy service on a co-primary basis. The band 2483.5 – 2500 MHz is also allocated worldwide on a co-primary basis to the fixed and mobile services. This same band is also allocated to the radiolocation service on a co-primary basis in regions 2 and 3, and in region 2 the band is also allocated on a co-primary basis to the radiodetermination satellite service. These bands are used, or planned to be used, by several mobile satellite PCS systems using non-GSO satellites.

Advantages
These bands are currently allocated to the MSS on a global basis.

Disadvantages

These bands are being used by MSS systems that have either just started or are about to start operation. The bands may only be available for satellite IMT-2000 in the longer term.

4.3.2.1.3 Frequency bands 2500 – 2520 / 2670 – 2690 MHz

These bands are allocated to the mobile-satellite service on a global basis. Both bands are also indicated on a worldwide co-primary basis to the fixed and mobile services, and in Regions 2 and 3 the bands are also allocated on a co-primary basis to the fixed-satellite service. The allocation of these bands to the mobile-satellite service is effective January 1, 2005/ (RR No. S5.414 and RR No. S5.419). These bands are used for different terrestrial applications in different countries.

Advantages

If other services can be phased out, these bands would contribute a significant portion of the required satellite IMT-2000 spectrum. These bands are already allocated to the MSS on a global basis and used or planned for use by the MSS in some countries. Making these bands available for satellite IMT-2000 would reduce the need for any additional MSS allocations to satisfy the satellite IMT-2000 spectrum requirements.

Disadvantages

In a number of countries this band is used for multi-point distribution systems (in some countries extensively) that have been deployed in urban as well as in rural areas. Licenses for this service have been recently granted for extended periods of up to 20 years. Phasing out of these services will therefore be very difficult for the foreseeable future and therefore may preclude its use for IMT-2000 in these countries.

4.3.2.1.4 Methods to satisfy the agenda item and their advantages and disadvantages

Studies conducted in the ITU-R indicate the incompatibility of the MSS (space-to-Earth) and ARNS/RNSS in any portion of the 1559 – 1567 MHz band. Not only do MSS signals have the potential to cause significant interference to ARNS/RNSS, but GNSS pseudolites and proposed new RNSS systems also have the potential to cause significant interference to the MSS (space-to-Earth). The RNSS is extensively used, and is continuing to undergo a tremendous expansion that drives further evolution. As a result of these factors, which have to be considered in conjunction with the many critical timing, positioning, and navigation uses of RNSS (including, but not limited to, aeronautical and maritime safety-of-life navigation), sharing of the 1559 – 1610 MHz band – including any portion of the segment at 1559 – 1567 MHz – with any co-frequency communication service is not recommended. Although studies were not carried out on every different type of RNSS receiver used in all the numerous applications of RNSS, it was nevertheless possible to conclude that sharing between ARNS/RNSS and MSS (space-to-Earth) is not feasible in any portion of the 1559 – 1567 MHz band.

4.3.2.1.5 Analysis of the results of studies

In the context of Resolution 218 (WRC-97), "prioritization" means assignment of the first available channel to the traffic within the AMS(R)S priority 1-6 in Article S44 or distress, urgency and safety communications of the GMDSS in accordance with RR No. S5.353A "Pre-emption", means terminating non-safety communications to establish the distress, urgency or safety communications. Within a network, under the definitions given above, prioritization and real-time preemption are deemed to be feasible and have been implemented in the AMS(R)S. However, an existing system has found that for its systems, the complexity and cost of implementing intra-system prioritization and preemption between different types of Mobile Earth Station (MES) standard designations is substantially greater than within one standard designation. Further studies are required to define the technical and operational requirements which will satisfy the needs of the AMS(R)S/GMDSS. Further studies are also required with respect to the implementation of a system that could use spectrum flexibly between different MSS networks. Possible impact on future spectrum requirements could result from the addition of redundant satellite system capability for the provision of AMS(R)S as well as the initiation of new types of data service.

4.3.2.2 Agenda Item 1.11

Agenda Item 1.11 is to consider constraints on existing allocations and to consider additional allocations on a worldwide basis for the non-GSO/MSS below 1 GHz, taking into account the results of ITU-R studies conducted in response to Resolutions 214 (Rev.WRC-97) and 219 (WRC-97). Resolution 214 (Rev.WRC-97) is entitled "Sharing studies relating to consideration of the allocation of bands below 1 GHz to the non-geostationary mobile-satellite service"

4.3.2.2.1 Summary of technical and operational studies relevant to sharing between existing non-GSO MSS systems and existing terrestrial services

A number of studies have been carried out since MSS allocations for non GSO satellite systems were first agreed at WARC-92. These have led to ITU-R Recommendations that indicate the sharing techniques, which are being used by those systems to share with each other and other co-primary services.

4.3.2.2.2 Analysis of the results of studies

Recommendation ITU-R M.1389 summarizes the techniques and Recommendations applied to existing MSS allocations. Many of these techniques are being employed in practice successfully.

4.3.2.2.3 Method to satisfy the agenda item

The constraints on existing allocations are reflected in the footnotes to the allocations, and in RR No. S9.11A. These have evolved to their present form since WARC-92. Administrations have accepted the sharing criteria between services of equal status. These constraints have served to provide a basis for implementing non-GSO MSS systems in these bands and at the same time provide protection to other space and terrestrial services. Therefore in respect of the constraints of the MSS in existing allocations below 1 GHz, no further modifications are needed.

4.3.2.2.4 Regulatory and procedural consideration

No modifications are required to the tables of criteria applicable to MSS allocations for use by noon-GSO systems below 1 GHz, as found in RR No. S9.11A, or to the footnotes containing constraints that apply to the pertinent allocations.

4.3.2.3 Agenda 1.12

To consider the progress of studies on sharing between feederlinks of non-geostationary-satellite networks in the mobile-satellite service and geostationary-satellite networks in the fixed-satellite service in the bands 19.3-19.7 GHz and 29.1-29.5 GHz taking into account Resolution 121 (Rev.WRC-97).

4.3.2.3.1 Summary of technical and operational studies

A draft new Rec. ITU-R S.[Doc.4/42 (Rev.1)] on mitigation techniques has been prepared. This Recommendation includes the topics of adaptive power control, high gain antennas, geographic isolation, site diversity and link balancing. In addition the Recommendation addresses coordination. The results of these studies are summarized below in Table 4.8

4.3.2.3.2   Method to satisfy the agenda item

The above section 4.3.2.3.1 is considered to cover the requirements of Resolution 121 (RC-97) and thus satisfies the agenda item. CPM-99 suggested that the Resolution 121 (WRC-97) could now be suppressed.

4.3.2.3.3   Regulatory and procedural consideration

The agenda item may be dealt with through ITU-R Recommendations rather than modification of the Radio Regulations.

4.3.2.4   Procedures

In addition to the allocation agenda items discussed above, a WRC will consider other regulations that impact the mobile satellite services. One of these matters is service definition. While the above has been discussing the mobile satellite service, in fact, this service is made up of three different mobile satellite services. These include land mobile satellite, maritime mobile satellite, and aeronautical mobile satellite. Their definitions are included in Article S1 of the Radio Regulations. Taken together they are considered the MSS, but they may also have their own individual allocations or one may be excluded, e.g., mobile satellite except aeronautical mobile satellite. In addition, a WRC will consider the allocation status of particular allocations as between Primary, and Secondary.

Finally, WRC will consider the procedures that govern the use of allocations. Different Mobile satellite allocations will be governed by different procedures. When the allocation is used by non-GSO systems, the Procedure is S9.11A, when it is used by GSO, the procedure is S9.7; when the MSS allocation requires special approval from countries of its coverage area then the procedure S9.27 applies. The texts of these procedures are in Appendix S5. All WRC's have on their agenda an item that provides for modification of these procedures. A WRC also considers regulations that may modify the sharing criteria regarding use of certain frequencies by MSS. Such criteria are typically recommended by WP-8D.

Table 4.8. Mobile-Satellite mitigation techniques

Technique	Description
Power Control	The use of adaptive uplink power control may be used to maintain system performance during times of increased levels of interference. Rec. ITU-R S.1255 recommends that networks employing adaptive uplink power control should transmit signals at the lowest possible power level to mitigate interference between GSO/FSS networks and feederlinks of non-GSO/MSS networks.
High gain antennas	Restricting the GSO earth station minimum antenna size to 1.0 m results in interference levels below the currently proposed aggregate criteria for the LEO A MSS network feederlinks. The interference level is generally higher for the MEO case than the LEO case. For a 1.8 m GSO earth station antenna the interference criteria are exceeded in all links.
Geographic isolation	As indicated in CPM Report to WRC-97, maintaining a minimum latitudinal separation of 2° between competing GSO/LEO earth stations is required to reduce the interference to acceptable levels. In the GSO/MEO case, separations greater than 2° in latitude (225 km) are required to reduce the interference to acceptable levels. Further, it has been shown that the geographic separation of the earth stations of two systems, combined with the use of high gain antennas, would be more effective in mitigating interference than either of the two techniques separately. Geographic isolation down to 60 km is possible with these techniques.
Site diversity	Site diversity is the use of an alternate earth station located far enough away from the primary site to provide sufficient antenna discrimination to maintain acceptable interference levels. Its use as an interference mitigation technique depends on the non-GSO MSS space station antenna beamwidth. For example, to be effective in mitigating interference between a LEO B feederlink network and a GSO/FSS network (GSO-13) in the 20/30 GHz bands, earth station diversity would require that the LEO B satellite antennas be so large as to be impractical.
Link balancing	The concept of link balancing refers to the design of non-GSO FSS links to reflect the need to mitigate the effects of uplink transmissions from GSO FSS earth stations. In the case of uplink transmissions from non-GSO MSS earth stations, the GSO receive signal is protected because of the distance involved. However, this is not the case for the uplink of the non-GSO MSS earth station vis-à-vis interfering transmissions from the GSO FSS. To "balance" the transmission environment, the non-GSO MSS earth station carries larger fixed uplink margins to protect itself from the GSO.
Coordination	To date the studies conducted by ITU-R indicate that geographic isolation provides the best solutions to coordination between non-GSO MSS feederlinks and the GSO FSS systems. Typically, there are relatively few non-GSO MSS feederlink earth stations in a system dispersed over a wide area. Within such an area, the non-GSO MSS earth station will require less spectrum than the GSO FSS, thus permitting the additional use of either frequency isolation and/or polarization isolation to achieve satisfactory coordination.
Satellite diversity	The use of satellite diversity has been considered as a mitigation technique to avoid main beam to main beam interference by switching traffic to an alternative satellite. This technique has a number of system design and network operational implications which network operators have to consider before implementing it. The constellation design is determined by how best to accommodate the service links and may not provide visibility statistics such that satellite diversity is possible.

4.3.3 Arrangements Concerning Global Mobile Personal Communications System (GMPCS)

4.3.3.1 Background

The GMPCS Arrangements comprise a number of elements related to the licensing, type-approval and registration of certain satellite earth stations. These elements are the activities that must be undertaken by the participants in the Arrangements which provide a regime for the unrestricted circulation of such satellite earth stations across the borders of participating member administrations. It is a new role for the ITU to serve as the depository for the GMPCS Arrangements. This task does not relate to any radio regulatory role currently being carried out by the ITU. The depository provisions in the GMPCS Arrangements is for the ITU Secretary-General to carry out the duties of the depository, indicated in the ITU Constitution (CS Article 1, No. 5). These encourage the ITU to undertake any tasks which will improve international communications. It was in pursuit of carrying out this role that the 1996 World Telecommunication Policy Forum (WTPF-96) established the concept of the GMPCS Arrangements. The ITU Council in 1997 and 1998 endorsed and provided the formal authorization for the ITU to become the depository for the GMPCS Arrangements. The GMPCS Arrangements as constructed cover any fixed or mobile satellite system, which provides telecommunication services directly to end users. The inclusion of fixed satellite systems in an Arrangement entitled "Mobile" came about very early in the development of the Arrangements when it became clear that personal communications provided by fixed satellites may require the same unrestricted circulation of satellite earth stations across borders as is required by mobile Earth stations. By the time that this realization was made, the acronym GMPCS had become well known, so no action was taken to change it.

4.3.3.2 Development of the GMPCS Arrangements

There have been three phases to the development of the GMPCS Arrangements.

4.3.3.2.1 Phase 1

The WTPF-96 decided that GMPCS Arrangements would be of considerable benefit to encourage the global extension of satellite communications. To achieve this goal to develop such Arrangements, the GMPCS Memorandum of Understanding (GMPCS-MoU) was established.

The purpose of the GMPCS MoU was that those Administrations and Sector Members which signed the MoU, agreed to work together to draft the GMPCS Arrangements. In December 1996, the ITU Secretary-General sent a letter to the Member States and the Sector Members inviting them to sign the GMPCS MoU and to participate in the GMPCS MoU Group. The GMPCS MoU Group held meetings from February 1997 to March 1998 when the GMPCS Arrangements and Implementation Procedures were completed.

4.3.3.2.2   Phase 2

In May 1998, the ITU Council authorised the Secretary-General to implement the GMPCS Arrangements on the basis of the full recovery of the costs to the ITU from the GMPCS industry participants in the Arrangements (ITU Council, Resolution 1116). In accordance with this decision, in July 1998, the Secretary-General sent a further letter to Member States and Sector Members inviting them to notify the ITU of their intention to implement these Arrangements. (When a sector member responds.) This is referred to as the "General Implementation" and it signals only an "intention" as a good will supporting gesture. The subsequent implementation is the system-specific implementation carried out in Phase 3.

4.3.3.2.3   Phase 3

All of the activities of phases 1 and 2 are good-will, cooperating activities led successfully to the implementation of System-Specific GMPCS Arrangements. The inauguration of System-Specific GMPCS Arrangements was initiated with a letter from the Secretary-General to Administrations in October 1998, inviting them to participate in the GMPCS Arrangements for the Iridium LLC GMPCS Satellite System. In the future, as each GMPCS system approaches its operational phase, it is expected that it will request the Secretary-General to initiate a system-specific implementation of the GMPCS Arrangements as done for the Iridium LLC system.

4.3.3.3   Features of the GMPCS Arrangements

In approximate chronological order, the steps of the system-specific GMPCS implementation are as follows:

a. The implementation is started when a GMPCS system operator sends a letter to the Secretary-General advising that (a) it is prepared to begin the global operation of its system, (b) that it has implemented the relevant provisions of the Arrangements and (c) requests the Secretary-General to undertake its system-specific implementation. In this letter, the GMPCS System Operator makes a commitment to uphold all of the provisions of the Arrangements including:

- renewal of its commitment to the voluntary principles in Opinion 2 of the Final Report of the WPTF-96;
- compliance with the relevant treaty provision and regulations of the ITU, and the national laws and regulations of each country in which its services are to be provided;
- the obtaining of the requisite licenses and other authorizations from the country it intends to serve;
- the taking of steps to inhibit the use of its system in any country that has not authorized its service;
- the provision of appropriate traffic data as required by applicable national legislation;
- the activating of only those GMPCS terminals authorized under the terms of the GMPCS Arrangements.

b. In concert with the above referenced letter from the GMPCS System Operator, the related GMPCS terminal manufacturers for a particular system, will also send a letter to the Secretary-General. In this letter the manufacturer describes the technical requirements that its terminals meet or exceed and makes a commitment to offer for use only those terminals for which this information has been deposited with the Secretary-General.

c. One or more Administrations advise the Secretary-General that they have, inter-alia, implemented the Arrangements, given type approval of the specific related GMPCS terminals, and have licensed or have initiated the national licensing process for the specific GMPCS system.

d. The Secretary-General advises those GMPCS terminal manufacturers which have met the terms of the Arrangements, for a specific system that they may affix the ITU GMPCS Registry Mark on their terminals and informs them of the guidelines for the use of the Mark. It should be emphasized that the ITU GMPCS Registry Mark shows only that the particular terminal to which it is affixed is registered with the ITU in accordance with the provisions of certain system specific Arrangements. While many of the provisions of the implementation procedure relate to type-approval, the ITU Mark is not a type-approval mark.

e. After the implementation steps defined in a-d above have been taken, the correspondence described in these steps is Annexed to a letter sent from the Secretary-General to all of the Member States inviting them to take the appropriate steps to implement the system-specific Arrangements for a particular GMPCS System, and requests them to notify the Secretary-General that the GMPCS terminals authorized for a particular system, are authorized to be used in their countries.

In many countries, the information attached to the letter from the Secretary-General will be sufficient to fulfil their national authorization requirements. In those countries where additional information is required the Administration may deal directly with the national service provider of a particular system.

f. Subsequently, the Secretary-General publishes a list and updates it from time to time, of the Administrations which are participating in the Arrangements of a particular system.

g. The Member States which have advised the Secretary-General that they will participate in a system-specific Arrangement, are required by the terms of the Arrangements to take appropriate action to ensure that the terminals of this system are allowed to enter their countries and can be used without constraint.

h. In the event that an Administration has not authorized a particular GMPCS system to operate in its country, that Administration is requested, under the provisions of the Arrangements, to allow the entry of such GMPCS terminals into its country, but too not permit their use the assurance that a terminal will not be used in such circumstances as provided for by the Arrangements whereby the GMPCS System Operator must be able to inhibit the use of its system in any country that has not authorized the service.

4.3.3.4   Summary

GMPCS Systems offer a variety of Radiocommunication services which require unrestricted transborder movement of their terminals in order to provide global service. The successful implementation of the GMPCS Arrangement through the cooperative efforts of Administrations the ITU GMPCS System Operators and terminal manufacturers will facilitate the early introduction of these new and innovative telecommunication services to the global communities, especially developing countries.

4.3.4   International Mobile Telecommunications – 2000 (IMT-2000)

This section describes the role of the Mobile Satellite Service in the context of IMT-2000. The IMT-2000 concept includes a terrestrial component, however, this aspect is not discussed in detail. Much of this material is taken from the Conference Preparatory Meeting (CPM) report dealing with this subject[CPM]. It was prepared by the participants of the ITU-R group, Task Group (TG) 8/1 charged with providing the technical basis for the WRC-2000 agenda item concerned with this topic (see also Section 4.2).

4.3.4.1   General Introduction to IMT-2000

IMT-2000 is the ITU vision of global mobile access in the 21st century. Scheduled to start service around the year 2000 subject to market considerations, IMT-2000 is an advanced mobile communications concept intended to provide telecommunications services on a worldwide scale regardless of location, network, or terminal used. Through integration of terrestrial mobile and mobile satellite systems, different types of wireless access will be provided globally, including services available through the fixed telecommunication networks and those specific to mobile users. IMT-2000 proposes a range of mobile terminal types, linking to terrestrial and/or satellite based networks, and the terminals may be designed for mobile or fixed use. The key features of IMT-2000 are:
   a. high degree of commonality of design worldwide;
   b. compatibility of services within IMT-2000 and with the fixed network;
   c. high quality;
   d. small terminals for worldwide use;
   e. worldwide roaming capability;
   f. capability for multimedia applications and a wide range of services (e.g., video-teleconferencing, high speed internet, speech and high rate data).
As the ITU provides the best means of ensuring that IMT-2000 will meet the telecommunications needs of all regions across the world, the ITU-R and ITU-T sectors are developing a set of interdependent Recommendations pursuant to IMT-2000. At WARC-92, 230 MHz of spectrum[1] was identified for FPLMTS via a provision (RR No. S5.388) in the Radio Regulations. IMT-2000 was formerly known as FPLMTS (Future Public Land Mobile Telecommunications Systems).

---

[1] This frequency spectrum was identified based on calculations that are now documented in Recommendation ITU-R M.687-2.

However, due to the tremendous growth in mobile communications since then, and the demand for wide-band multimedia capability, the expected demand for additional IMT-2000 spectrum after the initial deployment must be considered.

4.3.4.2   Satellite Component

In line with continuing technological advancements, more and more users will demand more and more capabilities from mobile services. Future mobile services must support not only speech but also a broad range of new telecommunication services that will serve a wide range of applications, as described in Rec. ITU-R M.816. Bearer services supporting applications such as multimedia, internet access, imaging and video conferencing will be needed in IMT-2000.   Forecasts of the spectrum requirements of the terrestrial and satellite components of IMT-2000 have been developed in Report ITU-R M.[IMT-SPEC]: "Spectrum requirements for IMT-2000". This Report uses expected market forecasts and assumptions on the operations of IMT-2000, for different environments and services, to determine the projected overall and additional spectrum requirements for IMT-2000.   For technical reasons (for instance propagation factors and terminal design), consideration of the additional spectrum requirements for IMT-2000 have focused on the same general frequency range as the original spectrum identified for IMT-2000; that is, below 3 GHz. For higher data rates, and where the user is stationary or nearly stationary, it may be desirable to utilize frequency bands above 3 GHz.

The Report of the CPM to WRC-95 indicated that a total allocation for MSS of between 2 x 75 MHz and 2 x 150 MHz would be required by 2005 (Chapter 2, part A.2 §3). Two years later, the Report of the CPM to WRC-97 indicated that 2 x 250 MHz would be necessary by 2010 (§4.2.6). A large number of satellite systems have been advance published by the ITU for the MSS frequency bands between 1 and 3 GHz; more than 150 systems in the 1.5/1.6 GHz bands, more than 50 systems in the 1.6/2.4 GHz bands, almost 100 systems in the 2 GHz bands and about 75 systems in the 2.5/2.6 GHz bands. Some of these systems have been filed in more than one of the bands. It is expected that some systems will not be implemented due to financial or other reasons. Nevertheless, the number of filings demonstrate the very large interest in providing MSS in the 1-3 GHz range.   Report ITU-R M.[IMT.SPEC] concluded that there is a forecasted need for mobile-satellite spectrum as shown in *Table 4.9*. In calculating this spectrum need, Report ITU-R m.[IMT-SPEC] applies the detailed methodology provided in Rec. ITU-R M.1391 to traffic estimates on the future demand for mobile satellite communications. The total MSS spectrum requirement is larger than that for the satellite component of IMT-2000 alone (see *Table 4.9*).

Table 4.9. Forecasted requirements for global mobile satellite spectrum, including IMT-2000 satellite component (MHz)

	Year 2005	Year 2010
IMT-2000 (Satellite Component)	2 x 31.5	2 x 67
Total MSS (including IMT-2000 satellite component)	2 x 123	2 x 145

Note: the figures in the table above represent the requirement in those geographic areas where the traffic is the highest.

In the various Regions, consideration should be given to the existing spectrum assigned for pre-IMT-2000 MSS services. Because the "Total MSS" spectrum calculation includes both pre-IMT-2000 and IMT-2000 services, a subtraction of the existing satellite spectrum allocated to the pre-IMT-2000 services must be performed to determine the additional satellite spectrum required for the IMT-2000 satellite component in the years 2005 and 2010. This subtraction has not been done in Report ITU-R m.[IMT-SPEC], due to the variations across Administrations of the spectrum allocated to pre-IMT-2000 satellite services.

4.3.4.3   Spectrum Vision for IMT-2000 Satellite Component

WRC-2000 agenda item 1.6.1 includes the satellite component since it is an integral part of IMT-2000, as indicated in Res. 212 (Rev. WRC-97). There must be a viable and operational satellite component to enable IMT-2000 to achieve the objectives stated in Rec. ITU-R M.687-2, particularly regarding true global coverage. The IMT-2000 terrestrial component will not be able to provide global coverage, and large geographic areas will remain without terrestrial coverage. The satellite component will provide IMT-2000 services in these geographic areas. Of the population within terrestrial coverage, some IMT-2000 customers will travel for both business and pleasure, to areas without terrestrial coverage. The high penetration expected for IMT-2000 means that some customers will want communications and high mobility anywhere and everywhere they travel. It is only through the combination of IMT-2000 terrestrial and satellite components that true global coverage can be accomplished.

It is also important to note that regions where high terrestrial coverage is anticipated may not be the driving force for determining spectrum requirements for the satellite component of IMT-2000. Rather, the areas of low terrestrial coverage are likely to be the main drivers for satellite IMT-2000 spectrum. Availability of global spectrum is particularly important for the satellite component. If global spectrum is not available, deployment costs could be prohibitive. The spectrum vision developed for the terrestrial component in *Figure 4.3* is also applicable to the IMT-2000 satellite component. The bands for the satellite component identified in Res. 212 (Rev. WRC-97) are 1980-2010 MHz and 2170-2200 MHz.

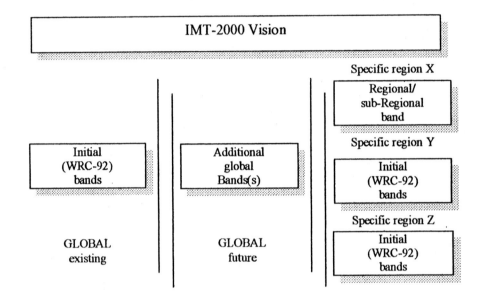

Figure 4.3. Spectrum Vision.

As the spectrum needed to support the IMT-2000 satellite component is discussed, it is useful to examine the situation of the 2 GHz MSS allocations identified by RR No.S5.388 and Res. 212 (Rev. WRC-97) that are intended to be used by the IMT-2000 satellite component. Res.212 (Rev.WRC-97) identifies the bands 1980-2010 MHz and 2170-2200 MHz for the satellite component of IMT-2000. However, the portion 1980-1990 MHz is not available in many countries in Region 2 for the mobile satellite service as per RR No. S5.389B. The mobile satellite service is also allocated in Region 2 to the 2010-2025 MHz and 2160-2170 MHz bands.

Non-IMT-2000 MSS systems are not precluded from using these allocations and are likely to do so, thereby reducing the amount of spectrum potentially available to support the IMT-2000 satellite component in this band. Sufficient frequency spectrum needs to be made available, in the year 2005-2010 time-frame, to allow for the possibility of competition. An important question for WRC-2000 to decide is whether the provision of additional satellite IMT-2000 spectrum comes from frequency bands already allocated to the MSS or if new allocations to the MSS should be made for such purposes. Consideration should be given to identifying existing MSS allocations between 1 and 3 GHz for satellite IMT-2000 applications. It is foreseen that most of the MSS bands between 1 and 3 GHz could be used for IMT-2000 in the longer term. However, the requirement for spectrum for other types of MSS systems also needs to be taken into account. An important factor when looking at spectrum for mobile satellite systems is the system replacement and evolution. Satellite systems equipment ages and must ultimately be replaced; however, replacement of this equipment is inherently a long-term process and can be difficult to accomplish. Therefore, use of currently occupied MSS spectrum for future satellite IMT-2000 systems will require substantial planning, investment and co-ordination with the current operators. It should be noted that, while future MSS systems will employ advanced techniques, resulting in greater spectrum efficiency than hitherto possible, this improvement has already been accounted for in the satellite spectrum requirements given above.

4.3.4.4   Summary of Technical and Operational Studies

4.3.4.4.1   List of relevant ITU-R Recommendations
Recommendations ITU-R M.1141-1 and M.1142-1 contain or address sharing studies that are relevant.

4.3.4.4.2   Additional Relevant Technical Information
Recommendations ITU-R F.1335 and M.1184 contain information on interference protection criteria and other information that, while not directly addressing potential IMT-2000 spectrum usage, may be useful for IMT-2000 sharing, technical and operational studies. Other ITU-R Recommendations containing useful information and characteristics for the mobile satellite service are:
ITU-R M.1091, M.1143-1, M.1183.

4.3.4.5   Analysis of the Results of Studies

4.3.4.5.1 Relevant ITU-R Sharing Studies

It is noted that agenda item 1.9 of the WRC-2000 deals specifically with the frequency band 1559-1610 MHz (Res.220 (WRC-97)) and 1675-1710 MHz (Res.213 (Rev.WRC-95)), which is addressed in section 2.2 of this report. These two bands are not considered further in this section.

4.3.4.5.2   Possible candidates for Additional Satellite IMT-2000 Bands

Consideration on candidates for additional IMT-2000 satellite bands has revealed that the following bands are potential candidates for IMT-2000 (see Table 4.10) below:

Table 4.10 . Possible candidate bands for the satellite component of IMT-2000

1525-1559/1626.5-1660.5 MHz
These bands are allocated to the mobile satellite service on a global basis. 1525-1535 MHz is also allocated on a co-primary basis to the space operation service (s-E). 1525-1530 MHz is also allocated in Regions 1 and 3 to the fixed service on a co-primary basis. 1660-1660.5 MHz is allocated on a worldwide basis to the radio astronomy service on a co-primary basis.
In the bands 1525-1559/1626.5-1660.5 MHz, MSS systems have been in operation for more than ten years. With an increasing number of new operators and increasing demand for MSS services, the bands are now rapidly approaching saturation, as is evident from the continuing regional multilateral co-ordination activities. Safety communications have priority in the bands 1530-1544/1626.5-1645.5 MHz as per RR No. S5.353A, and in the 1545-1555/1646.5-1656.5 MHz band, as per RR No. S5.357A. The bands 1544-1545/1645.5-1646.5 MHz are limited to distress and safety communications according to RR Article S31.
Advantages
Although the systems currently operating in these bands were state-of-the-art when launched, technological advancements will continue to lead to new systems being even more spectrum efficient. Hence, as the current MSS systems are eventually replaced by new systems, potentially including satellite IMT-2000 systems, such systems may be able to use these bands. These bands are allocated to the MSS on a global basis.
Disadvantages
The satellite component of IMT-2000 could only be implemented in these bands in the shorter term, taking into account spectrum congestion (see chapter 2, part A, §5).
Portions of the band 1525-1559 MHz and 16266.5-1660.5 MHz are subject to constraints. Deployment of IMT-2000 in these bands must take into account that usage in the sub-band 1544-1545/1645.5-1646.5 MHz is limited to distress and safety communications.

1610-1626.5/2483.5-2500 MHz

These bands are currently allocated on a global basis to the mobile-satellite service. The band 1610-1626.5 MHz is also allocated worldwide to the aeronautical radionavigation service on a co-primary basis. The band 1610-1626.5 MHz is in Region 2 also allocated on a co-primary basis to the radiodetermination satellite service. The sub-band 1610.6-1613.8 MHz is also allocated in all 3 Regions to the radio astronomy service on a co-primary basis. The band 2483.5-2500 MHz is also allocated worldwide on a co-primary basis to the fixed and mobile services. This same band is also allocated to the radiolocation service on a co-primary basis in Regions 2 and 3, and in Region 2 the band is also allocated on a co-primary basis to the radiodetermination satellite service.

These bands are used or planned to be used by several mobile satellite PCS systems using non-GSO satellites.

Advantages

These bands are currently allocated to the MSS on a global basis.

Disadvantages

These bands are being used by MSS systems that have either just started or are about to start operation. The bands may only be available for satellite IMT-2000 in the longer term.

2500-2520/2670-2690 MHz

The bands are allocated to the mobile-satellite service on a global basis. Both bands are also allocated on a worldwide co-primary basis to the fixed and mobile services, and in Regions 2 and 3 the bands are also allocated on a co-primary basis to the fixed-satellite service. The allocation of these bands to the mobile-satellite service is effective January 1, 2005. (RR No. S5.414 and RR No. S5.419). These bands are used for different terrestrial applications in different countries.

Advantages

If other services can be phased out, these bands would contribute a significant portion of the required satellite IMT-2000 spectrum. These bands are already allocated to the MSS on a global basis and used or planned for use by the MSS in some countries. Making these bands available for satellite IMT-2000 would reduce the need for any additional MSS allocations to satisfy the satellite IMT-2000 spectrum requirements.

Disadvantages

In a number of countries this band is used for multi-point distribution systems (in some countries extensively) that have been deployed in urban as well as in rural areas. Licenses for this service have been recently granted for extended periods of up to 20 years. Phasing out of these services will therefore be very difficult for the foreseeable future and therefore may preclude its use for IMT-2000 in these countries.

According to RR No. S5.403 and S5.420, the bands 2520-2535 MHz and 2655-2670 MHz may be used for MSS (except AMSS, see also RR No. S5.515A and S5.420A) for operation within national boundaries, and subject to agreement obtained under RR No.S9.21. These bands have been identified as possible candidate bands for the terrestrial component of IMT-2000, and are therefore not identified as suitable for satellite component. However, it is envisaged that these bands may be used for MSS in some areas, where the demand for satellite services is high.

The frequency bands 2010-2025 (E-s) MHz and 2160-2170 (s-E) MHz are allocated to the MSS, but only in Region 2. These bands are not available to the MSS in Regions 1 and 3. These bands are included in the spectrum identified in RR No. S5.388 for IMT-2000, and are adjacent to the global 2 GHz MSS allocations. However, these bands are not global MSS bands and their status as part of the satellite component of IMT-2000 is not addressed in Resolution 212 (Rev. WRC-97).

4.3.4.5.3 Methods to satisfy the Agenda Item and Their Advantages and Disadvantages

The spectrum requirements above might not be fulfilled in the existing MSS allocations and it is noted that further allocations will be considered by WRC-2000 to satisfy the agenda item 1.6.1,it is considered that there are 2 methods. These methods are described below.

Method 1

Additional global or Regional/sub-Regional bands could be listed through footnotes in article S5, with appropriate reference in Resolution 212 (Rev. WRC-97). The formulation of these footnotes for identification of additional spectrum for IMT-2000 could either be similar to RR No. S5.388 or adopt the format of footnotes such as RR No. S5.547, which is used for identification of the High Density Fixed Systems (HDFS) bands.

One alternative in this method to modify RR No. S5.388 to include the additional frequency bands; this footnote should have a link to the revised version of Res. 212 (Rev. WRC-97) or a new Resolution. Another alternative is to modify RR No. S5.388 to reflect the revised Res. 212 (Rev. WRC-97) in combination with the addition of a footnote specifying the additional bands, this footnote having a link to the updated version of Res. 212 (Rev. WRC-97) or a new Resolution. In both alternatives the different dates of availability for the additional IMT-2000 bands could be given in the revised version of Resolution 212 (Rev. WRC-97) or in the new Resolution.

Advantages

Having a footnote in the Table of Frequency Allocations for the IMT-2000 additional bands is consistent with the regulatory format adopted at WARC-92 to identify frequency bands for IMT-2000 in RR No. S5.388. The reference to IMT-2000 in the Table of Frequency Allocations highlights the use of these frequencies for IMT-2000 and could therefore facilitate the world-wide implementation of IMT-2000 and global roaming, without preventing administrations to allow other advanced mobile satellite applications in these bands if they wish.

Disadvantages

This different regulatory format for IMT-2000 systems compared to other mobile-satellite systems or radio services which are not footnoted in the Radio Regulations, could be misinterpreted as giving a different regulatory status to IMT-2000 compared with other systems.

Method 2

Additional global and Regional/sub-Regional bands could be listed in a new WRC Resolution or Recommendation, without any specific identification in a footnote to the Radio Regulations. Different dates of introduction for additional IMT-2000 bands can be given in this new WRC Resolution or Recommendation. Administrations may implement IMT-2000 systems in any appropriate frequency bands allocated to the mobile satellite service, therefore footnotes in Article S5 of the Radio Regulations are not essential. One alternative in this method is to maintain unchanged the RR No. S5.388 and Res. 212 (Rev. WRC-97) and to develop a new Resolution listing the additional satellite frequency bands and their dates of availability. Another alternative would be to suppress both RR No. S5.388 and Res. 212 (Rev. WRC-97), and to develop a new Resolution listing both the initial and additional satellite frequency bands for IMT-2000 and their dates of availability.

Advantages

Avoids misinterpretation of the regulatory status of IMT-2000 by disassociating any additional spectrum identification for IMT-2000 from the Table of Frequency Allocations.

Disadvantages

Listing the additional IMT-2000 frequency bands in a new WRC Resolution or Recommendation, without a corresponding reference in a footnote in the Table of Frequency Allocations gives a different regulatory format to the additional bands compared to the WARC-92 bands identified in RR No. S5.388. This might imply a different status between the existing spectrum identified at WARC-92 and the additional IMT-2000 bands. There is also the danger that the Resolution or Recommendation, if not referenced in a footnote, be disregarded by administrations. Furthermore, changing the regulatory format in this way might cause misconceptions that the world-wide implementation of IMT-2000 is becoming a lower priority than originally envisaged.

4.4   SUMMARY

Regulatory and spectrum requirements for mobile satellite systems to provide communications require consideration during the design phase and operation phase of these systems. These requirements should ensure a successful coordination of frequencies with other regional systems to guarantee satisfactory conditions exist to ensure operation without the presence of interference. Furthermore to ensure free circulation of mobile terminals, GMPCS Arrangements have been developed in regards to the licensing, type-approval and registration of certain satellite earth stations.

Future development of mobile satellite systems is most promising under the auspices of IMT-2000. IMT-2000 is the ITU vision of global mobile access in the 21st century, this advanced mobile communications concept intends to provide telecommunications services on a worldwide scale regardless of location, network, or terminal used. Through integration of terrestrial mobile and mobile satellite systems, different types of wireless access will be provided globally, including services available through the fixed telecommunication networks and those specific to mobile users. IMT-2000 proposes a range of mobile terminal types, linking to terrestrial and/or satellite based networks, and the terminals may be designed for mobile or fixed use.

ACKNOWLEDGEMENT

The authors wish to thank Ed DuCharme for his work on GMPCS aspects in this chapter.

REFERENCES

BOOKS
ITU-R Radio Regulations, Edition 1997.

GMPCS Reference book, December 1999,
http://www.itu.int/itudoc/itu-d/publicat/gmpcs.html.

REPORTS
CPM Report on technical, operational, and regulatory/procedural matters to be considered by the 2000 World Radiocommunication Conference.

ITU-R Recommendations, M series.

ERC Report 25, "Frequency range 29.7 MHz to 105 GHz and associated European table of frequency allocations and utilizations," Brussels, Brussels, June 1994 revised in Bonn, March 1995 and in Brugge, February 1998.

US proposals to CPM-99.

Chapter 5

Orbital Trades

Peter A. Swan[1] and Robert A. Peters[2]
[1]SouthWest Analytic,Inc,   [2] Stellar Solutions,Inc

5.1   INTRODUCTION

The ability to communicate anytime and anywhere across the global or regional customer base is key to commercial success. During the first generation of GMSS satellites the geosynchronous (GSO) orbit was chosen as the optimum for many reasons. INMARSAT is the result of an excellent space systems architecture approach during the 1970's, 80's, and 90's. However, global changes during the 90's drove satellite communications architectures to alternative concepts and orbits. Diverse factors such as customer mobility, customer diversity, smaller subscriber units, increasingly powerful spaceborne processors, commercial launches, global telecommunications deregulation, and global wealth enabled visionaries to pick alternatives for telecommunications constellations for the mobile user.

5.2   ARGUMENTS FOR NON-GSO ORBITS

The systems driver for GMSS architectures of the early to mid 90's was the time delay associated with signal path length. Traditional GSO one-way time delay was perceived as a significant deterrent to telecommunications satisfaction. To enable a global roaming customer base, systems architects realized that a quarter of a second delay in signals (approximately 35,000 Kms x 2 divided by the speed of light results in a quarter of a second time delay) was on the edge of acceptability for two-way conversations. This became especially annoying if the terrestrial link connecting the GSO relay was to go half way around the world - adding another tenth of a second to a one way link. This time delay was definitely within the realm of recognition during conversation and was classified as "annoying" by many consumers. These factors were especially visible when comparing orbital regimes for each new GMSS architecture. This chapter will quickly review orbital characteristics and summarize the choices that were, and are, being executed. There are three altitude regions and two orbital (conic) shapes being utilized for GMSS orbits (Figure 5.1). These are shown in Table 5.1, Orbital Regimes.

Table 5.1 Orbital Regimes

	LEO	MEO	GSO
Altitude (Km)	150-2,500	2,500-35,000	35,786
Shapes	Usually Circular	Circular and Elliptical	Circular
Inclinations	0-120 degrees	Usually 45-64 degrees	Zero degrees
# planes	3 to 6	2 to 4	Equatorial
Links	Direct downlink Relay satellite Cross-links or self relay	Direct Downlinks	Direct Downlinks
Typical GMSS Systems	Orbcomm, Vita Sat, Leo One, IRIDIUM® Globalstar	ICO-Global, Ellipso	Thuraya and AceS INMARSAT

SkyBridge[1], a 3rd generation mega-LEO satellite system, has chosen a non-GSO orbit for another reason; unavailability of sufficient bandwidth at GSO. While most broadband systems have been forced to the Ka band to get the desired spectrum, SkyBridge has proposed sharing spectrum with GSOs. Since all ground GSO terminals are directional and pointed at the geosynchronous orbit, then other antennas can point to non-GSO satellites. SkyBridge satellites will stop transmitting when their satellites are in a position where the tracking ground stations would interfere with GSO systems. The SkyBridge ground stations are re-pointed toward another satellite that is further from the GSO orbit. This greatly mitigates interference from SkyBridge ground terminals into GSO satellites and

[1] Not a GMSS, but a Broadband Multimedia Satellite System serving fixed antennas (FSS)

interference from SkyBridge satellites into GSO ground terminals. Non-equatorial orbit satellites have the potential of greatly increasing the bandwidth available.

5.2.1   Altitude Regimes

Traditional commercial communications satellite systems have been individual satellites at geoschronous altitude (35,786 kilometers) placed in longitudinal slots that were assigned by the International Telecommunications Union. The next step was to have a constellation of GSO slots filled with satellites for a single organization. An example is INMARSAT with its longitudinal locations or "slots" (98W, 25E, 64E, & 178E) to provide mobile communications to maritime users around the globe as well as aeronautical and land mobile units where authorized by the regional authorities. The second generation GMSS GSO systems, such as ACeS (80.5E, 118E, 123E, & 135E) and Thuraya (28.5E & 44E), are being inserted into orbital slots to provide regional coverage for their mobile customers.   These GSO orbits have been tremendously successful for commercial satellite builders and operators over the years and will continue to offer a very good location for large area coverage. This ability to have an orbital period match rotation of the Earth enables satellites to be viewed from Earth as "stationary". This orbital stability allows the use of simple receiving dishes without moving or tracking components.   In contrast to GSO orbits, Low Earth Orbits (LEO) satellites move rapidly over the Earth with short "in view" time for the customer because of orbital velocity (7.9 Km/sec) and small footprint of visibility on the surface of the Earth (approximately 4,000 Km. diameter for IRIDIUM).   As a result, LEO constellations depend largely upon wide angle (non-pointing) "stub" antennas. Communications have to be passed (handed over) from one beam to the next and from one satellite to the next automatically.   LEO constellations need large numbers of satellites (Globalstar – 48; IRIDIUM– 66) to enable continuous coverage around the globe. The third option, medium Earth orbit (MEO), is a compromise between GSO (major energy required - major regulatory issues) and LEO (with the complexity of many satellites with real-time dynamic connectivity needs). [See Table 5.1 above, Orbital Regimes]). ICO and New ICO have selected MEO orbits.

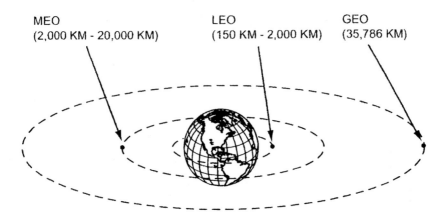

MEO   LEO   GEO
(2,000 KM - 20,000 KM)   (150 KM - 2,000 KM)   (35,786 KM)

Figure 5.1. Orbital Altitude Regimes

5.2.2   Low Earth Orbits

Low Earth orbit (LEO) architectures enable users to be closer to satellite antennas as the altitude of LEO satellites range from about 300 Kms to 2,000 Kms.   The lower limit is determined by atmospheric drag and the upper LEO limit by high radiation levels.   Low orbits have two important characteristics based upon their altitude. First, shorter distances mean less loss across the transmission path and less time delay in the signal propagation path.   Second, as the satellites are so close to the Earth, they have a very limited visibility and require a constellation of satellites to provide continuous coverage. The footprint of a single satellite might only cover a 4,000 Kms diameter circle on the earth's surface.   Therefore, to cover the surface of the Earth continuously, a basic requirement for real time communications, a constellation of satellites must have multiple overlapping coverage areas. For non-real time communications mission (similar to data messaging with Orbcomm) fewer satellites could be used to trade coverage time vs. satellite constellation costs. A real time communications system imposes a few more requirements on the architect such as ubiquitous coverage. This requires the orbital planner to place satellites in inclined orbits that are phased such that all potential customers' territories are covered at all times. The IRIDIUM system chose to cover the full Earth and resulted in a 66 satellite design to give global coverage (Figure 5.2). Globalstar chose to have a lower inclination and only cover the Earth between 60 degrees north and south latitudes. Their 48 satellite constellation always has one satellite, and much of the time has two satellites in view from a customer location. This creates a major advantage where blockage is a problem. These two choices reflect LEO architectures that need continuous coverage with varied inputs and trade spaces.

Another LEO architecture (Little LEOs) uses a small number of satellites (around 24), and smaller, less sophisticated/expensive satellites. A good example of this is Orbcomm with its mission of passing short messages. These space systems architecture choices by Orbcomm led to a LEO alternative while fulfilling the revenue objectives of many customers at low prices with small subscriber units. To achieve this, they deleted the continuous customer coverage criteria and used a store and forward technique. The process is to accept a signal when sent by the subscriber (after the satellite came into view), store the short message on the spacecraft, wait until the orbit places the satellite over a ground control station, and then dump the message to communications infrastructures on the ground. To send a message to a customer, the process is reversed and the time for message delivery is multiple minutes depending upon orbital dynamics instead of instantaneous as in IRIDIUM or Globalstar. The choice of a LEO system mandates many characteristics that are significant. When architects move into the LEO regime, they must consider the following parameters very carefully: orbital altitude, footprint (with minimum elevation angles), percentage of time satellite is in view, flexible pointing or staring beams, overlap in coverage from beams, orbital maintenance, cross links or ground infrastructure, and terrestrial command and control locations with backup.

5.2.3   Medium Earth Orbits

Another choice of orbits is medium Earth orbit (MEO) which encompasses all orbits above LEO and below GSO. Since this is a wide range of altitudes, the generally accepted regimes are above the lower Van Allen Belts (or 2,500 Kms) and up to about 20,000 Kms (half-GSO). The principle constellation utilizing this orbital region prior to GMSS architectures was the Global Positioning Satellite (GPS) System (20,232 Kms–circular–55 degree inclination) with 24 satellites. A driving systems requirement for the "semi-synchronous" orbit (12 hours) was a need to have at least four satellites visible to any location on the surface (or in the air) to derive the four unknown quantities in navigation (three dimensions + time). As the communications mission only requires one satellite at a time for real-time coverage, there are significant reductions in number of satellites to provide global coverage at MEO altitudes. The trade-off to this particular space systems driving requirement (# of satellites) at MEO, is the extra cost needed to launch into higher orbits, added radiation protection complexity and weight, and longer signal time delays. As a result, MEO choices for telecommunications are all closer to LEO than to the GPS constellation, (taking advantage of the range and coverage of LEO orbits).

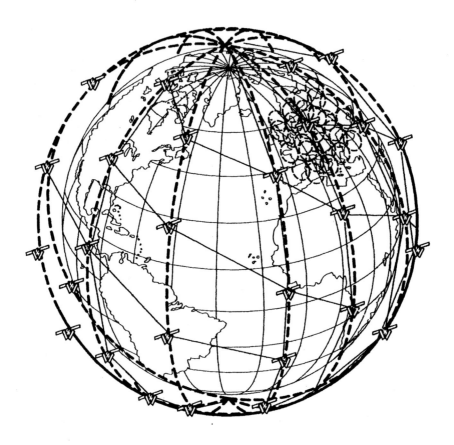

Figures 5.2. IRIDIUM coverage

The New-ICO, with its 12 satellites in circular orbits at 10,390 Kms altitude and 45 degrees inclination, is the principal example in a commercial MEO system. The choice of MEO by an architect leads to the following parameter investigations: visibility/coverage, number of satellites, radiation levels, hand-off patterns, beam coverage, gateway visibility, and telephony network connectivity.

5.2.4   Geosynchronous Orbits (GSO)

Geostationary orbits (GSO with zero inclination) allow the simplest orbital characteristics as they appear stationary from any point on the Earth. At an altitude of 35,786 Kms above the Earth's surface, a satellite takes one day to complete one orbit, so it is always at the same longitude and appears stationary from an Earthly perspective. The coverage from a geosynchronous satellite (for a 10 degree minimum elevation angle) is shown in Figure 5.3. While in theory, no energy is required to keep a satellite in orbit, in practice various forces must be countered, including solar pressure and uneven mass distribution of the Earth. These forces tend to move a satellite out of an exact geostationary orbit and require station keeping to maintain their designated orbit. The usual practice is to keep the satellite within a box of 0.5 degrees centered on their specified location. If the Earth station beam is smaller than about one degree, it may have to track the satellite's motion, adding to the cost of the antenna. Small televisions receive only (TVRO) antennas and VSAT (very small aperture terminals, usually used for low subscriber price GSO systems) do not need to track.

Figure 5.3 Coverage of GSO and LEO (780 km altitude) Satellites.

Even smaller terminals, such as those planned for CD Radio (satellite to car radio service), do not need to track.

An additional factor to be taken into consideration for GSO satellites is the radiation levels encountered during the geosynchronous transfer orbit (GTO) as it passes through both the inner and outer Van Allen radiation belts. This moderate level of radiation exposure usually rules out most commercial, non-radiation hardened parts. A significant increase in complexity and cost of launch is consistent with the effort to go from Earth to GSO. Transfer from a parking orbit in LEO, or direct insertion into the geosynchronous transfer orbit (GTO), requires much more energy and sophisticated upper stages not required by LEO systems.

5.2.5   Elliptical vs. Circular Orbits

In addition to various orbital altitudes, one GMSS system has chosen to alter the accepted practice of using circular orbits. Ellipso has proposed a constellation with elliptical orbits as the principal strength based upon customer needs. An elliptical orbit has both a perigee (lowest altitude point) and apogee (highest altitude point). The key is that there is a high velocity at perigee and low velocity at apogee in order to maintain the total energy constant of the orbit while trading altitude (potential energy) with velocity (kinetic energy). This leads to rapid motion over the Earth at perigee and slower motion at apogee. This enables the satellite to "dwell" above an area over the Earth for a longer period than if it were in a circular orbit. The Ellipso constellation is unique in that it has three orbital planes that are different. The Concordia plane has a circular orbit with seven satellites with zero inclination at an altitude of 8,060 Kms . This 4.6 hour orbit enables the equatorial and mid-latitude regions (approximately 60° north to 60° south latitudes) to be covered by these MEO satellites as they stay in view of a customer on the equator for much longer times than a LEO satellite. This orbit covers the dense population areas of India and Indonesia. The second type of orbit is elliptical with a three-hour orbit at an inclination of 63.4 degrees (necessary to keep the perigee from walking north). The perigees of these two Borealis planes are at 673 Kms , while the apogees are at 7,515 Kms . This provides a majority of satellites in the northern hemisphere, as the time spent in the southern hemisphere is less because of the high velocity perigee passage. This results in a long dwell over the high population areas of the United States, Europe and Russia at its apogee, or slow velocity passage. In addition, the phasing of these orbits emphasizes viewing angles across Europe and the United States during the high usage times of early morning and late afternoon. This is accomplished with two elliptical planes of approximately 180 degrees difference in ascending nodes. The characteristics of the Ellipso constellation proposal have many strengths and a few concerns such as increased costs to the higher orbits and the multiple transitions through the Van Allen radiation belts per day.

5.3   CONSTELLATION DESIGN

The traditional GMSS method to achieve customer satisfaction was to put a satellite at GSO and coordinate the payload and antenna beams so that they support the user. The minimum number of satellites in this type of constellation is usually given as three (120 degrees of longitude separations); this gives up coverage at the two poles with some overlap at the horizon so the angle to the subscriber unit is not horizontal requiring long paths of atmospheric transmission. INMARSAT grew to greater than three satellites to provide better service to ships at sea with multiple beams staring at required locations across the ocean and some land masses. A minimum constellation size of one is the smallest for a GMSS architecture and relates directly to customer needs and regional initial revenue start-up service from GSO. As the orbit decreases in altitude, the number of satellites required to support uninterrupted service increases. Many considerations must come into architectural trades at this point. As shown in the previous sections of this chapter (Figure 5.3), the coverage per satellite on the Earth gets smaller. This ground circle is called the ground coverage and relates to the altitude and the incident angle at the horizon. If the coverage is to the horizon, the elevation angle (or grazing angle) at the horizon is zero. As this puts the signal through the longest atmospheric path, the zero grazing angle is not used except for physical measurements. The minimum grazing angle depends upon frequency, data rate, bit error rate and power at both transmitter and receiver. Usually, the angle is no less than 8 degrees and is often up to 12 and 15 degrees. Figure 5.4 shows the orientation of elevation angles: zero, 45 and 90 degrees.

A second factor in the evaluation of satellites for GMSS is the number of satellites required at any one time within view of a subscriber unit. The IRIDIUM system requires one satellite to be in view of each spot on the surface of the Earth at all times. This is true at the equator where the constellation is spread out between the planes; however, as the satellites travel north and south, they "bunch" together as the planes merge toward the poles (Figure 5.5). This provides multiple satellites at all times above/below certain latitudes.

The Orbcomm system is even more efficient in that it only requires that one satellite be over a single point on the ground periodically. This means that there is no continuous coverage requirement and fewer satellites can be used to cover the globe. Customers know that the system might, or might not, be above them at any specific time so the system is a non-real time messaging system, by constellation design. In contrast, Globalstar has designed their constellation of satellites to have multiple satellites in view of a subscriber - especially if they are in the mid latitudes (i.e. the high density affluent countries of Europe and North America). This design overlaps coverage routinely and provides an ability to do power adjustments on signals to/from its handsets. Once again, a design aspect impacts the constellation set-up. The top end of coverage is the Global Positioning Satellite System that provides navigation to the world.

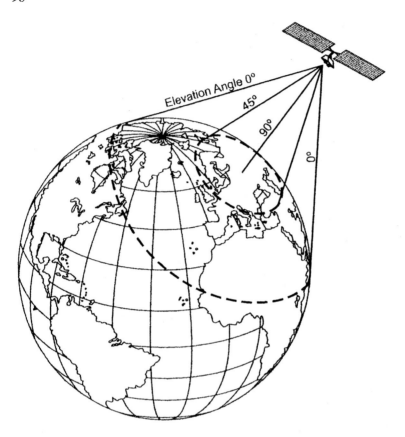

Figure 5.4 Elevation Angle

    Its requirement is to be able to have four satellites in view at any one time. They are required to have 21 satellites consistently with three spares on orbit for a total of 24 satellites at an altitude of 20,000 Kms. In contrast, the New-ICO satellite system requires only one satellite above the subscriber at any one time and is in a MEO orbit with only 10 satellites required for operations (with two spares on orbit). This range of satellite nodes in a GMSS constellation goes from one, when dealing with a regional GSO, to 66 for IRIDIUM. The largest constellation registered at the FCC over the last few years was the Teledesic constellation of 840 operational satellites. This has since been reduced to 120 satellites - still a large number. In addition, ground coverage can be optimized for certain locations by providing a non-circular orbit. John Draim, of Ellipso, has established a set of orbits that optimize coverage for GMSS constellations using two elliptical planes and a circular plane in equatorial orbit. This design led to 17 satellites for the Ellipso constellation.

Several major factors need to be evaluated when looking at constellation designs. The basic rule is: higher altitudes require fewer satellites; however, higher altitudes mean higher RF losses, more radiation and higher launch costs. In addition, GSO satellites usually have longer life spans (15 years). With this in mind, the following items must be studied over the development of a constellation concept.

Constellation Pattern: This issue is one of type of coverage best suited for the services to be provided. Some factors involved in this topic are: coverage at the poles (GSO, no; LEO polar, yes; LEO with no or slight inclination, no), the number satellites that must be in view at any one time, and the needed repeat of ground traces (good for connectivity with ground infrastructure). There are many "solutions" to this puzzle, but there are two well-known ones:

Walker Orbits: They provide the smallest number of satellites for continuous coverage of the surface of the Earth. This leads to a very consistent spread of satellites around the globe implying that all spots on the Earth are equal in value. Five satellites is the minimum coverage number for a LEO/MEO constellation. The Walker Delta pattern usually has common inclinations, phase angles within planes, and an accepted notation of "t" satellites in "p" orbital planes.

Adams Rider Orbits: The orbits in this category all cross at two nodes: northern and southern hemispheric passages. This enables constellation operators to plan for a routine footprint overlapping and cell turn-off as satellites go further north and south where multiple satellite coverage occurs. As all planes cross the equator evenly spaced around the world (except at the seam between north and south passing planes) and pass close together at two nodes, the scheduling of cell shutdown and satellite maintenance can be easily accomplished.

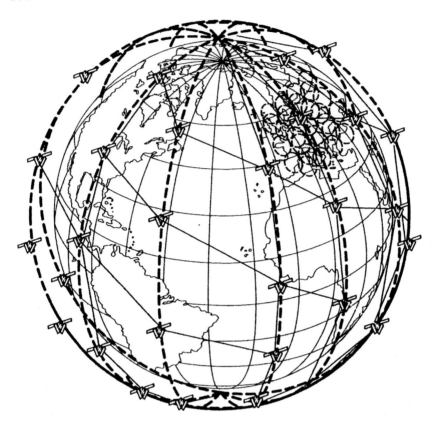

Figure 5.5 Polar Planes Converging Orbits

Number of Satellites: This is truly one of the key elements of constellation design as it directly impacts the cost of constellations. In addition, it remains a major factor throughout the constellation's life with replacement and operations for each satellite. The trade-off is between number of nodes/satellites and the cost of each. There is a phenomenon occurring in this GMSS telecommunications revolution that is significant -- the production line manufacturing of satellites instead of hand crafting them. Globalstar, IRIDIUM, and Orbcomm all had production lines that "mass-produced" identical satellites at greatly reduced cost.

Altitude: As the number of satellites goes down and the altitude goes up, many factors change; connectivity to subscriber units with path length, atmospheric penetration, and beam diversity. In addition, as you go up in altitude, radiation becomes severe above about 1500. The Van Allen belts are global and continuous; so, if your altitude is higher than 1500   the satellite goes through at least the lower belts on the way to mission orbit. A GSO orbit is above both the major belts of radiation and, in addition to traversing them on its way to orbit, a satellite then does not have as strong an Earth's magnetic field to protect it from solar radiation and intergalactic particles. A third factor is the cost of attaining altitude. All satellites during their launch phase must get to LEO altitude and then proceed to their mission orbits. If that is in LEO, many less expensive launch approaches work. To go to MEO, GSO or Elliptical orbits, upper stages must be used at an increase in cost, complexity, and risk. Some other critical factors are coverage provided by each satellite (number of satellites required for desired coverage), orbital period, communication link path loss, and system cost (radiation issues plus launch costs).

Minimum Elevation Angle: This has been discussed earlier but is very significant. As the minimum elevation angle increases, the coverage on the ground decreases (Figure 5.3) and more satellites are required to cover a global user base. If the path goes to the horizon, or zero minimum elevation angle, the path losses in the atmosphere and the probability of blockage increase greatly. These trades are significant in the overall constellation design process.

Propagation Paths: The mission of GMSS is customer satisfaction through a subscriber unit that is reasonable to carry and handle. This means that the propagation of the signal has to be designed carefully to ensure the power on the ground is not too much (for safety concerns), the antenna on the ground is not too large (cost and handling), and the attenuation throughout the path is not too great. In addition to the design aspects of physical components, the selection of the frequency is critical to attenuation levels and the ability to close the communications link.

Link Handoffs/Handovers: Systems architecture must accommodate satellite-to-satellite handovers. If, as is usually the case, the satellite has multiple beams, then the hardware and architecture must handle beam-to-beam handovers within the satellite as well. While a non-processing satellite can be used if the number of beams is not too large, onboard processing is required to handle the increase in complexity when the number of beams becomes large. Globalstar uses non-processing satellites (each Globalstar satellite has multiple beams and does not use inter-satellite links) and handovers are handled entirely by ground equipment. IRIDIUM uses processing satellites (each IRIDIUM satellite has 48 beams and has inter-satellite links) and handovers are handled by the satellites. Either way, a LEO system is far more complex to operate than a GEO system because of the periodic handovers and the large number of satellites required to provide continuous coverage.

Number of Orbital Planes: This parameter usually falls out of the trades for the constellation pattern; however, the criticality of the number is important to ensure that systems architects maintain a watch on this parameter. An extra plane usually means several more satellites, hence a higher constellation price. Fewer planes mean that it is harder to provide critical ground coverage with higher minimum elevation angles. This trade is sensitive to the orbit altitude and usually produces plateaus of answers.

Inclination of the orbits: For the normal constellation, the answer is consistency between each of the planes. Remember, inclination of orbits gives the constellation designer flexibility for coverage requirements. The Ellipso constellation uses elliptical orbits at the critical inclination of 63.4 degrees. This emphasizes coverage of the northern hemisphere while the Globalstar system uses circular orbits that are at 48 degrees inclination. This emphasizes both of the temperate zones (north and south of the equator) and allows multiple satellite coverage over areas of large gross national product (GNP) countries. IRIDIUM uses polar orbits that provide consistent coverage over the globe (includes the poles for rare activities such as scientific or exploration expeditions). IRIDIUM's stated intent was to have a consistent constellation that would be easy to manage with inclinations of 86.4 degrees and plenty of overlap at the poles for time to conduct routine maintenance of the satellites during light loading windows.

Eccentricity: The normal design of a constellation is to use a circular orbit. However, as it is very hard to ensure circularity (exactly zero eccentricity) in a physical orbit, most orbits have some small eccentricity. IRIDIUM even designed its planes to have varying eccentricities to ensure acceptable miss distances at the poles when the planes crossed. The ultimate communications orbit is a GSO with circular orbits. If the owner of a geosynchronous location desires to have multiple satellites at a node (to handle such issues as more transponders or separate frequencies) then some eccentricity is again required to avoid collisions or frequency interference. In addition, the Ellipso constellation was designed to use eccentricity as a major factor. John Draim, Director of Constellation Design for Ellipso and Virtual Geosatellite LLC, has over the last 10 years developed multiple options for a mix of elliptical and circular orbits to optimize coverage around the globe. One intriguing aspect of this approach is that he has patented the orbits (US patent # 5,582,367) and has thus gained protection for this uniqueness.

Disposal Strategy: The US government has mandated upon its DoD and civil satellite systems the requirement to have a disposal plan for each satellite that it places in orbit. This would include disposal into the upper atmosphere for low Earth orbiting satellites, raising to a graveyard orbit at GSO, and placing in a graveyard orbit in-between GSO and LEO. A significant issue in the arena today is where are the graveyard orbits for each of the high LEO and MEO orbiting constellations. One person's mission orbit could be another's graveyard orbit. Table 5.2 shows some systems proposed for the region of 1350 Kms altitude to 1500 Kms . Each of the shells, described by an altitude/radius, contains a number of satellites to accomplish the communications mission. Globalstar is already at 1450 Kms , circular, with 48 degrees of inclination. In addition to the shells listed below, the elliptical orbits that have perigees lower than the altitudes of these satellites, will be going through the shells each time they go to perigee and raise themselves to apogee. An interesting conundrum is that there is no "traffic cop" to ensure that satellites do not collide with one another at orbital velocities.

There are many other less critical factors that go into the design of a constellation for a GMSS. Some of them are: launch alternatives (launch vehicles as well as locations), collision probability within the constellation and with other constellations, location flexibility for orbital maintenance, cross-link sensitivity, number of ground stations for command and control of a constellation, number of ground stations for telephony mission, customer coverage such as regional or global, time of day coverage, and radio frequency interference with GSO satellites.

5.4   COMPARISON OF ORBITS

There are many different tradeoffs that must be accomplished prior to the determination of orbital parameters. There are many that directly effect customer relationships and many, while still important, that are secondary. Table 5.3 illustrates some general characteristics of five GMSS second generation systems.

Table 5.2. Constellation Shell Conflicts (as of 1999)

Name	Owner	Type*/Freq	#Sat	Altitude Kms/ Inclination Deg
Teledesic	Teledesic, LLC	Mega - Ka Band	288	1375 / 85
Globalstar	Globalstar LP	Big - L/S Band	48	1414 / 52
Skybridge	Skybridge LP	Mega-KuBand	80	1457 / 55
Ellipso-Borealis	Ellipsat (MCHI)	Big - L Band	10	673x7515 / 63
Ellipso-Concordia	Ellipsat (MCHI)	Big - L Band	7	8060 / 0
Signal	KOSS	Big - L Band	48	1500 / 74
Hughes NET	Hughes	Mega - Ku Band	70	1490
Gonets D	Smolsat	Little-UHF	36	1400 / 82.6
Gonets R	Smolsat	Little-S/L Band	45	1400 / 82.6
E-Sat	E-Sat (Echostar)	Little-VHF/UHF	6	1260 / 100

The selection of orbits is only one aspect of the architecture and probably one of the last to be finalized. The comparison of GMSS constellations based upon orbits is multiple-dimensional and only a part of the total required space systems architectural tradeoffs. Table 5.4 shows the major trade items associated with altitude, but does not lead to any conclusions as it is only one part of the equation.

Table 5.3. GMSS Constellation Characteristics

	Altitude (Km)	# of Satellites	# of Planes	Eccentricity degrees	Inclination degrees
Orbcomm	785	21	3	Zero	45
Globalstar	1450	48	8	Zero	48
IRIDIUM	780	66	6	Zero	86.4
Thuraya	GSO	2	1	Zero	Zero
ACeS	GSO	2	1	Zero	Zero

Table 5.4. Altitude Characteristics

	Advantages	Disadvantages
LEO Low 150-500 Km	Closest to the subscriber units. Lowest cost to orbit	Most satellites for continuous coverage - drag
LEO Med 500-1,000	Inexpensive to orbit -Good visibility – Low radiation	Needs many satellites
LEO High 1000-2,000	Excellent coverage Less satellites needed	More expensive to orbit. Radiation effect design
MEO 20000-30,000 Km	Least satellites required Longer dwell times	In radiation environment. More expensive to orbit
Elliptical Apogee< 10,000 Km	Focused coverage Excellent dwell times	Unique operational aspects Expensive to orbit
Elliptical with high Apogee	Tremendous coverage of northern tier area Long dwell times (12 hour repeating traces)	Very expensive to orbit Regional per set of sat's Large ground antenna High rad. environment
GSO	24 hour dwell time. High data rate to large staring antennas. Low data rates to small mobile apertures. Least number of satellites Simple antenna.	Greatest path delays. Very expensive to orbit. Regional per set of satellites

A trade-off leading to the best system to accomplish a desired service has to be conducted at the systems architectural level. Table 5.5 shows the major advantages of LEO and GEO orbits.

Table 5.5. Comparison of Advantages of High and Low orbits

Geosynchronous Satellites	Low Earth Orbit Satellites
High path loss	Much smaller path loss (20 to 28 dB less)
Stationary antennas	Omni antennas with tracking for high data rates
System capacity difficult to expand because of limited orbital slots	System capacity easy to expand through-out orbital shell
No hand-offs required	Beam to beam hand-offs Satellite to satellite hand-offs
Low Doppler, slowly changing	High Doppler, rapidly changing
Satellite covers 40% of earth's surface, 3 satellites need for 100% coverage(no poles)	>40 satellites needed for continuous coverage
Constant path delay	Rapidly varying path delay
Long path delay (1/4 sec round trip)	Short path delay 10 to 30 msec. Round trip

Low Earth orbit (LEO) satellites have six major advantages over geosynchronous (GSO) satellites.

(1) They have typically 20 dB (or greater) less path loss because of the distance with greater sensitivity to small uplink signals.
(2) The launch costs are significantly reduced when only going hundreds of kilometres.
(3) LEO systems have substantially less path delay than GEO systems. This is an important consideration for some digital protocols such as internet connectivity.

(4) Orbits are less than about 1500 km, damaging space radiation is less and commercial parts, rather than radiation hardened electronics, can be used.

(5) If medium gain ground antennas are used, the same spectrum can be reused many times by a LEO system, as discussed earlier in this chapter. As there is often little spectrum available in the geosynchronous arc, this may be the only way of operating a new space system.

(6) LEOs have effectively a regulatory advantage over GSOs. Whereas each GSO satellite is independently registered and coordinated, a LEO constellation is registered and receives its license for all of the satellites. A LEO gets its own orbital "shell" for its satellites when it receives its license to operate. Providing interference remains within any coordination agreements, the number of satellites in the shell can be increased at the behest of the constellation owner. In addition, they can share the same frequency as a GEO satellite in systems that use pointing antennas (as proposed by SkyBridge).

Some disadvantages also exist for LEO/MEO constellations.

(1) Multiple satellites are required to provide continuous coverage over any one region. The number of satellites for a complete constellation can range from 40 to several hundred. Globalstar has 48 satellites with eight on orbit spares, IRIDIUM has 66 satellites with six spares, and Teledesic has proposed 288 satellites with a large number of spares. Usually, the whole constellation must be operational before service can start, which translates into substantial up-front investment prior to revenue initiation.

(2) A second problem with LEO satellites is that satellite to satellite handovers are required for continuous operations. As one satellite moves out of visibility and a new satellite moves into visibility, the link must be handed off to the new satellite. Simultaneously with the subscriber handoff, the gateway link must also be transferred. This change takes place so rapidly that it is invisible to the subscriber. Satellite to satellite handoffs are required every 5 to 20 minutes. As altitudes increase, Earth coverage velocity decreases and the handoffs are simplified. The ultimate is the GSO orbit with zero motion on the Earth; therefore, there are no velocity handoff issues. If a LEO satellite has multiple static beams, then a beam to beam handover is required as the satellite passes overhead. Consider the following hypothetical example. A LEO satellite passes directly overhead, is visible for 10 minutes, and has 25

beams. The footprint diameter has about 7 beams as shown in Figure 5.6. A beam to beam handover is required every 10/7 or 1.4 minutes. This handover requires a protocol, satellite to terminal communications, a "reservation" of resources in the new beam to accept the handover, and represents an increased chance of link breakage. This reduces satellite capacity and increases satellite cost. There must be sufficient overlap of the beams to allow time for the handover, which also decreases capacity. Fortunately, the architecture to handle these handovers has already been developed by the terrestrial cellular industry. In a cellular system, users must be handed off to other base stations as they move. In a LEO system, the user is (relatively) constant; but, the "base station" is moving.

(3) Beam pointing complicates systems architectures of GMSS constellations. A GEO satellite with small beams must be pointed very accurately to provide the desired coverage. A LEO satellite does not have to point as accurately as the user can be handed off to the optimum beam. Pointing is important only to insure proper overlap between satellites.

(4) A fourth complication is that there is a large and changing Doppler component to the frequency. Doppler is the frequency change due to relative velocity between a source and receiver. The relative velocity is high when the LEO is at a low elevation angle and low when it is overhead. The relative motion also results in a changing path delay that must be corrected for in TDMA systems. While GSOs have some relative velocity and hence Doppler, it is several orders of magnitude lower and changes slowly. However, the amount of Doppler is deterministic and predictable.

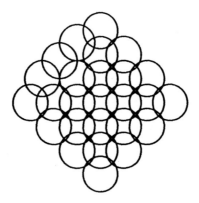

Figure 5.6. Hypothetical 27 beam LEO satellite

(5) If a LEO ground terminal antenna uses a directional antenna (an antenna with gain greater than an omni antenna), it needs to track the satellite across the orbital motion. An unbroken link during handover requires a second antenna to establish a make-before-break connection with the ascending satellite, making the ground segment cost substantially more than for a single GSO antenna. For low capacity systems, a GSO antenna does not have to track. Throughout the telecommunications industry there are several efforts to develop a low cost phased array ground terminal. This allows electronic steering with two different beams (to two different satellites) from the same antenna.

(6) A sixth concern for LEO constellations is the number of satellites between 24 and 288. This means that their production and launch costs must be reduced greatly over that of single satellite GSO systems to be competitive. LEOs became feasible when the concept of an assembly line was applied to satellite production. This challenging concept, new production line, has been applied. Factories have been developed and satellites produced to ensure reduced costs per unit. As an example, the IRIDIUM production line was delivering a satellite every 4.3 days at the peak of production with a total assembly time per spacecraft of 14 days total.

5.5   TRENDS IN ORBITS

There are many forces in the GMSS arena that will enable creativity and change. Some of the significant events and forces are: regulatory crisis at the ITU and the FCC, bandwidth scarcity at GSO, commercial launch industry over-capacity, spacecraft processors with significantly more power, global telecom revolution, mobile population expecting to be in contact at all times, and an 'Internetted' public wanting more (connectivity, speed, data rates, coverage, etc.). These forces and events will drive GMSS systems architects to create more flexibility in their systems - to include orbit selections. Some of the future selections of orbits will be to a larger extent toward elliptical and MEO/high LEO.

REFERENCES

BOOKS
Wertz, James R. and Wiley J. Larson (editors). Space Mission Analysis and Design. Third Edition. 1999. Boston: Kluwer Academic Publishers.
REPORTS
Satellite Tool Kit, Satellite Systems Analysis Software. Malvern, Pa. Analytical Graphics, Inc. 2000.

JOURNAL/SYMPOSIA ARTICLES
Drain, John and G. Helman, D. Castiel. "The Ellipso Mobile Personal Communications System; Its Development History and Current Status." (paper # IAF-97-M.3.03) 48th International Astronautical Congress. Turin, Italy: October 1997.

Drain, John and C. Davidson, D. Castiel. "Evolution of the ELLIOSO and ELLLIPSO 2G GMPCS Systems." (paper # IAF-99-M.4.02) 50th International Astronautical Congress. Amsterdam. October 1999.

Drain, John and C. Davidson, D. Castiel. "VIRGO, A "Virtual-GEO" Elliptic Orbit Fixed Satellite Communication System for the 21st Century." (paper # IAF-99-M.4.03) 50th International Astronautical Congress. Amsterdam, The Netherlands. October 1999.

Maine, Kris, Carrie L. Devieux, Jr. and Peter A. Swan "Overview of IRIDIUM® Satellite Network," IEEE Western Communications Satellite Systems Conference, San Francisco, California, November 1995.

Swan, Peter. A. and T. Garrison, M. Ince, and J. Pizzicaroli. "Systems Engineering Trades for the IRIDIUM® Constellation," JOURNAL OF SPACECRAFT AND ROCKETS, Vol. 34., No. 5, Sept-Oct 1997.

Swan, Peter. A. "IRIDIUM® Constellation Dynamics - The Systems Engineering Trades," (IAF-95 -U.2.04), 46th Congress of the International Astronautical Federation, Oslo, Norway, 5 October 1995.

Swan, Peter. A. "Manufacturing Technologies, the "Key" to a 66 Small Satellite System," (IAF-94-U.3.475), 45th Congress of the International Astronautical Federation, Jerusalem, Israel, 9 October 1994.

Swan, Peter. A. "77 to 66 - The IRIDIUM® Improvement," (IAF-93-M.4.339), 44th Congress of the International Astronautical Federation, Graz, Austria, 16 October 1993.

Chapter 6

Propagation Considerations

Randy L. Turcotte
Tempe, Arizona

6.1 INTRODUCTION

In the development and design of communication systems, the communication engineer is responsible for the optimization of system performance at a reasonable cost. Channel characteristics play a major role in determining system performance. Unfortunately, the engineer has no control over these characteristics. Physical properties dictated channel characteristics and these characteristics must be considered when optimizing system performance.

The channel influences the selection of many system parameters. These include modulation type, channel coding, antenna selection and transmit power to name just a few. It is imperative that communication system engineers have an understanding of the important channel characteristics, as well as, a way of modeling channel effects. The development of satellite communication systems requires the system engineer to make a variety of architectural decisions. A channel model is an essential engineering tool that allows engineers to explore the system trades necessary to optimize system performance at an affordable cost.

A significant body of work exists on the characterization of terrestrial based land mobile channels. Books by Jakes (1974) and Lee (1982) are recommended sources. While many of the basic concepts and descriptions developed for the terrestrial channel are closely related to the land mobile satellite system (LMSS) channel, differences do exist, requiring specific satellite channel models.

For example, terrestrial systems seldom have a line-of-sight path available between the mobile user and the base station. The utilization of scattered signal allows communications to be carried out. On the other hand, transmit power is generally not a limitation for a terrestrial base station while satellite systems are typically power limited. It is prohibitively expensive to generate high transmit power on a satellite platform or to install large antenna structures on a platform. To save cost, satellite systems must operate at a lower fade margin than terrestrial systems. On the bright side, most satellite systems have a line-of-sight signal component available and this must be accounted for in the satellite channel model.

Ray tracing or geometric models have the appeal of being able to relate a physical scenario to a statistical result. In these models, multiple surfaces that scatter electromagnetic energy (scatterers) are placed randomly according to well-defined criteria. The model is then used to predict various signal statistics. Unfortunately, real world scenarios can be extremely complicated, making these models limited to simplified situations with only a small number of scatterers.

Probability distribution channel models are based on the premise that the satellite channel can be classified as either being shadowed or unshadowed. In each case, the total received signal is comprised of various types of signal components, with each signal component having a probabilistic description. The probability density function (pdf) for the total received signal is derived from the density functions of the individual signal components. These models lend themselves well to system level issues where the probability descriptions are used for both analysis and simulation. Matching model parameters to parameters derived from measured data further enhances the models.

Each type of model has its place in the analysis, development and characterization of LMSS. Because of their ease of use and wide acceptance, this chapter focuses on the probability distribution type of channel model.

6.2   SATELLITE GEOMETRY AND RECEIVED SIGNAL COMPONENTS

Figure 6.1 below provides a simplified depiction of the LMSS channel showing the primary signal components of interest. The direct or unshadowed component arrives via a line-of-site path from the satellite to the mobile user. The atmospheric effects of absorption, scintillation and Faraday rotation influence this component.

A diffuse component may be created by reflections from multiple scatterers near the vehicle. The energy from the satellite is reflected toward the vehicle from each of the many scatterers. The waves arrive at the receiver with random polarization, amplitudes and phases. Each contribution is delayed by varying amounts of time, depending upon differences in path lengths.

A shadowed signal component occurs when trees or foliage obstruct the line-of-site path between the satellite and the vehicle. As the signal passes through the foliage, it can be absorbed, diffracted, scattered or a combination of all three effects.

A phase coherent specular component is caused by reflections within the first Fresnel[1] zone of the vehicle. Because the reflection is phase coherent, such reflections can cause deep fades when the amplitude of the reflected wave is on the order of the direct component's amplitude. Fortunately, the transmissions of most LMSS employ circularly polarized waves. When the angle of incidence of a circularly polarized wave is above the Brewster[2] angle, the polarization of the reflected wave will be oppositely polarized relative to the incident wave. For L-band transmissions, the Brewster angle typically is between 6° and 27°. With elevation angles for most satellite systems lying above 20°, the specular component is most likely to be oppositely polarized with respect the direct component. The combination of polarization discrimination and antenna pattern discrimination allow LMSS propagation models to ignore the contribution due to the specular component.

The atmospheric effects of Faraday[3] rotation, scintillation[4] and absorption affect all components that make up the received signal from the satellite. As most LMSS systems employ circular polarization, the effect of Faraday rotation is irrelevant. However, the losses due to scintillation and absorption should be considered.

---

[1]   As a wave front encounters obstacles, it will be diffracted. This diffraction results in an electromagnetic interference pattern, with both constructive and destructive interference. The first Fresnel zone is the change in the position of a scattering or diffracting element in the direction perpendicular to the line-of-sight path required to produce a phase shift of 180 degrees relative to the signal propagating in the line-of-sight or direct path.

[2]   When an arbitrarily polarized wave is incident at the Brewster angle, the polarization of the reflected electric field will be parallel to the boundary. This causes circularly polarized incident waves to be reflected with elliptical polarization.

[3]   Faraday rotation is a rotation of the plane of polarization caused by the interaction of the electromagnetic wave as it passes through the atmosphere.

[4]   Variations in electron density of the ionosphere will cause rapid variations, scintillation of the signal amplitude, phase and direction of arrival.

Figure 6.1   Simplified LMSS channel diagram

Most LMSS channel models and link budgets account for the losses due to scintillation and absorption as a lump sum degradation by allocating a small amount of extra margin. To see the degree to which scintillation and absorption affect L-band transmissions, one can consider predicted values for each. A plot of gaseous absorption at L-band frequencies (1.5 GHz) is shown in Figure 6.2. Gaseous losses at an elevation angle of 10° are about 0.3 dB, decreasing rapidly at higher angles. For most cases, the absorption due to gaseous effects may well be an order of magnitude less than the uncertainty due to other propagation effect. Allowing for a loss of approximately 0.2 dB should be adequate to account for gaseous losses for L-band transmissions.

Scintillation causes rapid short-term fluctuations in the received signal strength, which occur at a rate much faster than the symbol rate of the received signal. These rapid variations in signal amplitude tend to be averaged out during the demodulation and detection of the signal. As long

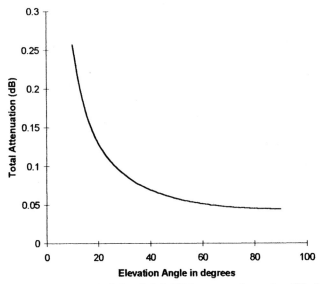

Figure 6.2 L-band (1.5 GHz) gaseous absorption (Flock, 1987)

as the range over which the signal amplitude varies is not too great compared to the average received signal level, degradations due to scintillation can be accounted for as fixed degradation. An example of the degree of L-band scintillation is found in Figure 6.3. From Figure 6.3 it is seen that degradations due to scintillation affect L-band propagation to about the same degree as gaseous absorption. At an elevation angle as low as 10°, scintillation for a 1- percent time of fade exceedence, results in a decrease in received signal level of approximately 0.3 dB. Most channel models combine the losses due to gaseous absorption and scintillation as a constant loss factor by allowing for a small increase in fade margin. A combined loss factor of approximately 0.5 dB should be adequate when accounting for the combination of gaseous and scintillation losses. This 0.5 dB factor is simply a rule of thumb and will vary depending upon desired reliability.

Other L-band atmospheric losses, such as those due to rain and clouds are an order of magnitude smaller than the gaseous absorption and scintillation and can simply be ignored.

Several channel models based upon probability distributions have been proposed (Loo, 1985; Lutz, 1991; Smith and Stutzman, 1986). Each of these models judiciously combines the primary received signal components to account for various channel conditions. Before addressing the channel model itself, a review of the fundamental probability distribution functions for each of the received signal components is included.

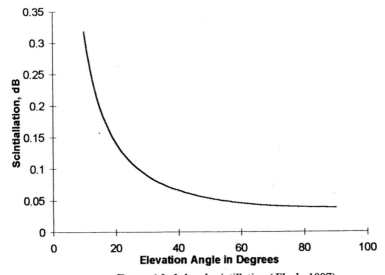

Figure 6.3 L-band scintillation (Flock, 1987)

6.3 PROBABILITY DISTRIBUTIONS FOR FADING MODELS

The Rayleigh, Rician and Lognormal probability distribution functions form the basis for the probability distribution class of channel models. These integrated channel models are developed by appropriately combining these functions to form a complete probabilistic description of the LMSS channel. Before considering the LMSS models, it is informative to review the channels described by each individual distribution function.

Figure 6.1 shows some of the possible signal components observable at the receiver. The direct component or non-faded component arrives at the receiver attenuated and time delayed[5] but undistorted. If a simple continuous tone was transmitted, the received signal *x(t)* is described as:

---

[5] A linear phase shift across the frequency band of the signal would also be considered undistorted.

$$u(t) = A\cos(2\pi f_0 t + \theta) \tag{6.1}$$

where A is the signal amplitude, f_0 the carrier frequency and θ an arbitrary phase angle. The individual probabilistic channel models are derived by considering how the channel changes $u(t)$.

6.3.1 Rayleigh Fading – Diffuse Signal Components

A Rayleigh fading channel is the result of the antenna receiving a large[6] number of reflected waves from nearby objects. These reflected waves arrive at the receiver with various amplitudes, different Doppler frequency shifts and at different arrival times. A simple signal such as $u(t)$ is transmitted. The received signal is conveniently expressed in terms of quadrature components of the carrier:

$$r(t) = x(t)\cos(2\pi f_0 t) - y(t)\sin(2\pi f_0 t) = z(t)\cos(2\pi f_0 t + \theta_0) \tag{6.2}$$

where

$$x(t) = \left[\sum_{n=1}^{N} A_n \cos(2\pi f_n t + \theta_n)\right] \qquad y(t) = \left[\sum_{n=1}^{N} A_n \sin(2\pi f_n t + \theta_n)\right] \tag{6.3}$$

$$z(t) = \sqrt{x(t)^2 + y(t)^2} \qquad\qquad \theta_0 = \tan^{-1}[y(t)/x(t)]$$

Here A_n, f_n and θ_n are the amplitude, Doppler frequency shift and phase of the nth reflected wave. The components $x(t)$ and $y(t)$ are each considered to be the sum of N statistically independent scattered waves. When the number of scattered waves is large enough these terms are approximated as zero mean narrowband Gaussian processes, each with the variance

$$\sigma^2 = \sum_{n=1}^{N} A_n^2. \tag{6.4}$$

Following the density functions for $z(t)$ and θ_0 are found to be

---

[6] In this case, large can be as few as 6.

$$f_z(z) = \frac{z}{\sigma^2} \exp\left\{-\frac{z^2}{2\sigma^2}\right\} \quad z(t) \geq 0 \qquad \theta_0 = \frac{2}{2\pi} \qquad -\pi < \theta_0 \leq \pi \qquad (6.5)$$

The density function for envelope z(t) is the well-known Rayleigh distribution. It is also noted that the envelope *z(t)* and the phase θ_0 are statistically independent random variables.

It should be kept in mind that this model is an approximation; for example, it implies that the mean signal power is constant whereas it actually undergoes slow variation as the receiver moves through its environment. Nonetheless, the model is applicable for a wide range of problems.

6.3.2   Rician Fading – Direct Signal Component with Diffuse and/or Multipath Components

The Rayleigh channel described in Section 6.3.1 is a rather severe channel. The received signal is comprised solely of multiple scattered components, resulting in a Rayleigh distributed signal envelope with a uniformly distributed phase. Such channels exist in some terrestrial environments. In the case of mobile satellite systems, a direct signal component is typically present as part of the total received signal. When a direct signal component is present with a large number of scattered components, the received signal is the sum of a direct signal component and a Rayleigh faded component. In this case the received signal can be written as

$$r(t) = z(t)\cos(2\pi t f_0 t + \theta_0) + A\cos(2\pi f_0 t + \phi) \qquad (6.6)$$

The first part of Equation (6.6) is the Rayleigh faded signal described in Section 6.3.1 and the second half is the direct signal component with known amplitude *A*, carrier frequency f_0 and phase ϕ. As in the previous section, r(t) can be written in terms of its quadrature signal components:

$$r(t) = [A\cos(\phi) + x(t)]\cos(2\pi tf_0 t) - [A\sin(\phi) + y(t)]\sin(2\pi f_0) \quad (6.7)$$

where, the envelope of the received signal becomes

$$v(t) = \left\{[A\cos(\phi) + x(t)]^2 + [A\sin(\phi) + y(t)]^2\right\}^{1/2} = \left\{v_c^2(t) + v_s^2(t)\right\}^{1/2} \quad (6.8)$$

For any given value of ϕ both quadrature components $v_c(t)$ and $v_s(t)$ are uncorrelated Gaussian variables, and therefore independent. For any given value of ϕ, the mean and variance of the quadrature components are simply:

$$E\{v_c(t)\} = A\cos(\phi) \qquad\qquad E\{v_s(t)\} = A\sin(\phi)$$

$$(6.9)$$

$$\mathrm{var}\{v_c(t)\} = \sigma^2 \qquad\qquad \mathrm{var}\{v_s(t)\} = \sigma^2$$

The joint density of the quadrature components $v_c(t)$ and $v_s(t)$ is the conditional Gaussian density:

$$f_{v_c v_s}(v_c, v_s \mid \phi) = \frac{1}{2\pi\sigma^2}\exp\left\{-\frac{1}{2\sigma^2}\left[(v_c - A\cos\phi)^2 + (v_s - A\sin\phi)^2\right]\right\} \quad (6.10)$$

The density function of the envelope $r(t)$ is found to be Rician:

$$f_V(v) = \frac{v}{\sigma^2}\exp\left\{-\frac{(v^2 + A^2)}{2\sigma^2}\right\}I_0\left(\frac{vA}{\sigma^2}\right) \quad (6.11)$$

The direct signal component of Equation (6.6) is the signal of interest. The Rayleigh component can be viewed as noise-like interference. Indeed, because the Rayleigh component can be expressed in terms of a quadrature Gaussian signal, it is in many ways analogous to additive Gaussian noise. In a manner similar to a signal-to-noise ratio, it is useful to define the parameter, K.

The average received power due to the Rayleigh and direct signal components is given by:

$$P_{Ray} = \sigma^2 \qquad\qquad P_{dir} = \frac{A^2}{2} \qquad (6.12)$$

Considering the power in the direct component to be the desired signal and the power in the Rayleigh component to be noise, the K value is defined as

$$K = \frac{P_{dir}}{P_{Ray}} = \frac{A^2}{2\sigma^2} \tag{6.13}$$

Several ways of normalizing Equation (6.11) are available. One convenient form is to define the variable:

$$w = \frac{v}{\sigma} \tag{6.14}$$

The Rician density function of Equation (6.11) may now be expressed as

$$f_W(w) = \frac{w}{\sigma} \exp\left\{-\frac{w^2}{2} + K\right\} I_0\left(w\sqrt{2K}\right) \tag{6.15}$$

As the power in the Rayleigh component increases, the value of K decreases. By allowing K to go to zero, it is easy to see that the Rician density of equation (6.15) reduces to the Rayleigh density of Equation (6.5). Of course, as power in the Rayleigh component tends toward zero, K goes to infinity and the envelope will no longer be random.

The phase angle is described by its density function conditioned on ϕ or, assuming an arbitrary, but known value, $\phi = 0$ the density function of an angle α becomes:

$$f_A(\alpha) = \frac{\exp(-K)}{2\pi} + \frac{K\cos(\alpha)}{\sqrt{4\pi}} \exp\left[-K\sin^2(\alpha)\right]\left\{1 + erf\left[K\cos(\alpha)\right]\right\} \tag{6.16}$$

where α is defined as

$$\alpha = \tan^{-1}\left\{\frac{v_s}{v_c}\right\} \tag{6.17}$$

The error function is defined as:

$$erf(x) = \frac{2}{\sqrt{\pi}} \int_0^x e^{-y^2} dy \qquad (6.18)$$

6.3.3 Lognormal Fading – Shadowing

The Lognormal probability distribution arises when one considers the product of independent random variables, much in the way that the Gaussian distribution arises from the sum of independent random variables. To see the analogy, consider the product of independent positive random variables

$$U = \prod_{i=1}^{N} U_i \qquad (6.19)$$

Define a new random variable Y as the logarithm of U. Taking the logarithm of both sides of Equation (6.19), the random variable Y is:

$$Y = \ln(U) = \sum_{i=1}^{N} \ln(U_i) = \sum_{i=1}^{N} y_i \qquad (6.20)$$

When Y is the sum of a large number of variables, its distribution will become Gaussian. By transforming the Gaussian distributed random variable Y back to the original random variable U, one shows that the density function of U is lognormal.

$$f_U(u) = \frac{1}{u\sqrt{2\pi\sigma^2}} \exp\left\{\frac{[\ln(u)-m]^2}{2\sigma^2}\right\} \qquad (6.21)$$

Where the mean and variance of U are respectively:

$$m = E\{Y\} = E\{\ln(U)\} \qquad \sigma^2 = \text{var}\{Y\} = \text{var}\{\ln(U)\} \qquad (6.22)$$

Shadowing occurs when the transmission from the satellite passes through some type of foliage, trees for example. The absorption and scattering of the signal passing through a tree can be modeled by a sequence of attenuation coefficients corresponding to the branches and leaves of the foliage. The received signal amplitude can be modeled as:

$$U = A \prod_{i=1}^{N} c_i \qquad\qquad (6.23)$$

Where U is the received signal amplitude, A is the signal amplitude if the path was line-of-site and the c_i's are the random attenuation coefficients. Comparing Equation (6.23) and Equation (6.19) it is easy to see why the shadowed signal amplitude is modeled by a lognormal distribution.

6.4   CHANNEL DYNAMICS

The probability distribution functions described in Section 6.3 characterize the received signal at an instant in time. Channel conditions, particularly those of a channel associated with a mobile user are expected to vary over time. The motion of the mobile user results in a shift and possible spectral spreading of the received signal know as Doppler spread.

6.4.1   Doppler Frequency Spread and Time – Selective Fading

If channel conditions result in the presence of a large number of multipath signals distributed about the receiver with no direct signal component, the envelope of the received signal is Rayleigh distributed. With the addition of a direct signal component, the received signal envelope becomes Rician distributed. If the direct component is shadowed, the received signal envelope has a Lognormal distribution. These distributions characterize the received signal at an instant in time.

Motion of the receiver and/or the satellite causes an additional effect known as Doppler spread. The power spectrum of the received signal is spread over a finite bandwidth. To better understand how Doppler spread comes about, consider a pure sinusoidal tone such as that described in Equation (6.1) being transmitted to a moving vehicle. Let the vehicle traveling at a speed v. A given reflected signal, one of many, arrives at the receiver from an angle β, where β will be distributed according to the probability density function $p(\beta)$[7] When the transmitted frequency is f_0 the frequency of a particular wave arriving at the receiver will be undergo a frequency shift dependent upon the velocity of the vehicle and the angle of arrival of the reflected wave. The frequency of a particular received signal is

---

A change in notation will be used here for clarity. The probability density function will be denoted p(•) to avoid confusion with frequency f.

$$f = f_0\left[1 \pm \frac{v}{c}\cos(\beta)\right] = f_0 \pm f_m\cos(\beta) \tag{6.24}$$

Where f_m is the maximum Doppler shift and c is the speed of light. The maximum Doppler shift occurs when the vehicle is moving directly toward or away from the source of the reflected wave ($\beta = 0, \pi$). When a large number of reflected wave arrive at the receiver, the spectrum of the received signal will consist of a set of spectral lines, each occurring at random frequency in the range $f_0 \pm f_m$. The probability that a spectral line will fall within a differential frequency band is given by the product of the probability density function of the received signal frequency, $p_F(f)$, and the frequency differential df. Because the spread in Doppler frequency is a function of the angle of arrival β, the frequency content in the differential df can be obtained from the probability density function of the angle of arrival, β as

$$p_F(f)|df| = \{p_\angle(\beta) + p_\angle(\beta)\}|d\beta| \tag{6.25}$$

where term df is the differential of Equation (6.24). The power spectrum may be found by first solving for $p_F(f)$, solving for the time autocorrelation of the channel impulse response and taking the Fourier transform of the result. If the reflected waves arrive at the receiver from uniformly distributed angles the power spectrum at the output of an omni-directional antenna is

$$S(f) = \frac{3\sigma}{2\pi}\sqrt{1 - \left(\frac{f - f_0}{f_m}\right)^2} \tag{6.26}$$

where σ is the mean signal power received by the antenna. The spectrum described in Equation (6.26) is the spectrum found when a large number of scattered signal components arrival at an omni-antenna from a set of uniformly distributed scatterers. In the LMSS channel, a line-of-site path commonly exists. If a line-of-site path exists and the transmitter were to transmits a CW tone, the received power spectrum will be a combination of the CW tone with its carrier frequency shifted by the Doppler frequency and the power spectrum described in Equation (6.26).

Of course in an actual system, the transmitted signal would be a modulated carrier. As long as the baud rate of the modulation is significantly smaller than the carrier frequency f_0, the received signal can be approximated as a frequency shifted version of the transmitted signal, plus the Doppler spectrum.

6.4.1.1 Time - Selective Fading

As stated above, the Doppler spectrum is the Fourier transform of the channel autocorrelation function. The channel autocorrelation function is simply the correlation of the channel's impulse response with itself. The result is a second-order characterization of the channel, which provides information of how the channel's impulse response varies [8] over time. The coherence time of the channel represents the time difference over which the channel impulse response remains strongly correlated. The larger the channel coherence time, the less the Doppler spread. A channel with a large Doppler spread indicates that the channel characteristics are varying rapidly with time, leading to a fast or time-selective fade. To account for these changes, the distribution functions of Section 6.3 are combined with the dynamic channel characteristics described in Section 6.4.1 to model the LMSS channel.

6.4.1.2 Frequency – Selective Fading

A time-selective fading channel is a consequence of the channel characteristics varying rapidly in time. In an analogous manner, it is possible for the frequency characteristics of the channel to vary over the bandwidth of the transmitted signal. Such a situation leads to a frequency-selective channel. As Doppler spread characterizes a time-selective fading channel, delay spread characterizes a frequency-selective fading channel.

Consider a multi-path channel where a number of time-shifted and scaled versions of the transmitter signal arrive at the receiver. This distribution of path delays is known as the delay spread of the channel. Delay spread causes the transfer function of the channel to vary over frequency, resulting in frequency-selective fading.

In a manner analogous to Doppler spread, delay spread is characterized in terms of coherence bandwidth. The coherence bandwidth is the frequency range over which the channel characteristics remain correlated. The larger the coherence bandwidth of the channel, the smaller its delay spread.

6.5 THE PROBABILITY DISTRIBUTION BASED LAND MOBILE SATELLITE CHANNEL MODEL

The communication system engineer is interested in understanding propagation characteristics in order to make informed decisions in the

---

[8] Or equivalently, the channel transfer function.

system development, design and operation of a communication system. Channel characteristics impact the selection of modulation format, coding scheme, antenna complexity, transmit power and system operation. One of the most useful tools the communication system engineer has is the channel model. A channel model must be accurate enough to model the characteristics of interest, while not becoming too complex and cumbersome to use.

The basic satellite channel geometry, the fundamental probability distributions and the effects of channel dynamics were summarized in the previous sections of this chapter. These pieces to the puzzle are now brought together to create a probability distribution based, LMSS channel model. Several varieties of probability distribution based channel models have been proposed (Loo, 1985; Lutz, 1991; Smith and Stutzman, 1986). Because of limited space and similarities between the models, we shall focus on the model proposed by Loo (Loo, 1985). The interested reader is encouraged to consider and review the other models.

6.5.1 The Loo Model

A probability distribution based channel model that has shown reasonable agreement to measured data is due to Loo. The primary statistical description of this model is the result of judiciously combining the Rayleigh, Rician and Lognormal channel distributions described in Section 6.3. Loo considered the satellite link of Figure 6.1 and hypothesized that as the signal passes through foliage, it is attenuated and scattered. The line-of-site signal component that passes through foliage undergoes lognormal fading. It is also assumed that a large number of multipath components are present, which result in a Rayleigh distributed component. These two components are assumed to be correlated and additive. The resulting probability distribution is developed in Loo (1985) and its development is summarized here.

The received signal is the sum of the two signal components, one lognormal and the other Rayleigh distributed. To find the resulting distribution the sum of the two components it is convenient to first hold the lognormal component fixed. Under this condition, the sum is comprised of Rayleigh component and a fixed component. As described in Section 6.3.2, the result is a signal with a Rician distributed envelope. The condition of holding the Lognormally distributed component fixed is removed by applying the law of Total Probability. The condition of holding the lognormal component fixed is removed by multiplying conditional Rician density by the lognormal density of the line-of-sight signal component and integrating. The resulting probability density of the received envelope is an

integral expression, which in general, must be evaluated numerically. The
density of the envelope is derived in Loo (1985). The result is repeated
here in . P_m is the average power due to multipath scattering and σ_s^2 and m_s
are the variance and mean due to shadowing.

$$f_R(r)=\frac{r}{\sqrt{2\pi\sigma_s^2}P_m}\int_0^\infty\frac{1}{x}I_0\left(rx/P_m\right)\exp\left\{-\frac{(\ln(x)-m_s)^2}{2\sigma_s^2}-\frac{(r^2+x^2)}{P_m}\right\}dx \qquad (6.27)$$

When the value of the envelope is much greater than the standard deviation
of the multipath (Rayleigh) process, the envelope is approximately
lognormal. When the envelope is much less than the standard deviation of
the multipath, the envelope is approximately Rayleigh distributed. Loo has
determined parameter values that provide a good match to various
environments. These are summarized in Table 6.1.

Table 6.1 Channel Model Parameters

	Light Shadow	Average Shadow	Heavy Shadow
P_m	0.158	0.126	0.0631
m_s	0.115	-0.115	-3.91
σ_s	0.115	0.161	0.806

6.5.1.1   Secondary Statistics

In Equation (6.27), Loo provides a statistical description of the envelope of
the received signal. It is equally important for the communication system
engineer to understand the fading dynamics of the channel. Loo provides
this information by calculating the secondary statistics of level crossing rate
(LCR) and average fade duration (AFD). The LCR is the expected value of
the rate at which the signal envelope exceeds a defined threshold[9].

---

[9] That is, the envelope crosses the threshold with a positive slope.

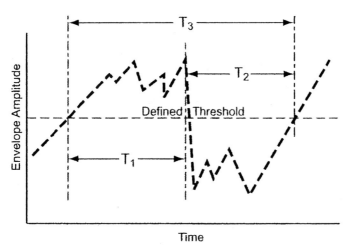

Figure 6.4 Simplified diagram of a signal envelope as a function of time.

The average fade duration is the expected value of the time the signal envelope is less than a defined threshold.

Consider the three intervals T_1, T_2 and T_3 as defined in Figure 6.4. The jagged line in the figure represents the random fluctuations of the signal envelope. During the interval T_1 the value of the envelope exceeds the threshold. The expected value of T_1 is then the average value of the non-fade duration. Similarly, the expected value of T_2 is the average fade duration, that is the average time that the envelope is below the threshold.
Now T_3 is equal to the sum T_1 and T_2. It is also the interval between times where the signal envelope crosses the threshold in a positive direction, so the expected value of T_3 is also equal to the reciprocal of the LCR. Equivalently, the reciprocal of the LCR is equal to the sum of the average fade and non-fade duration.

Knowledge of these statistics is important in the selection of various system attributes. For example, the AFD will impact the selection of the modulation format; channel coding scheme and the interleaver depth.
The derivation of the LCR expression is somewhat length, can be found (Loo,1985) and is based on previous work (Rice, 1944, 1945, 1948; and Jakes , 1974). The essential points of the development are summarized here.
The starting point for the derivation is the general expression for the LCR of a wide sense stationary random process developed by Rice (Rice, 1944, 1945, 1948).

$$LCR = \int_0^\infty r' f_{Rr'}(R, r')dr' \qquad (6.28)$$

Where the ' indicates the derivative with respect to time, r is the envelope and $f_{Rr'}$ is the joint probability density function of the envelope and its time derivative. Recall that the Loo model combines the effects of a Rayleigh multipath process and a lognormal shadowing process. The time rate of change of the envelope due to multipath is denoted by x' and the rate of change of the envelope due to the shadowing process is denoted by y'. The rate of change of the envelope is itself shown to be $r' = x' + y'$, where x' and y' are correlated Gaussian processes with correlation coefficient $\rho_{x'y'}$. The LCR is derived to be

$$LCR = \frac{1}{\sqrt{2\pi(1-\rho_{x'y'}^2)}} \frac{\sigma_m^2\sqrt{\sigma_m^2 + 2\rho_{x'y'}\sigma_m\sigma_s + \sigma_s^2}}{\sigma_m^2(1-\rho_{x'y'}^2) + 4\rho_{x'y'}\sigma_m\sigma_s} f_R(r) \qquad (6.29)$$

Where σ_m^2 is the variance of x' and σ_s^2 is the variance of y'. In many instances, it is convenient to normalize the LCR with respect to maximum Doppler frequency, making the result independent of velocity. The normalized LCR can be shown to be

$$LCR_N = \sqrt{2\pi(1-\rho_{x'y'}^2)}\sigma_{mo}^2 \frac{\sqrt{\sigma_{mo}^2 + 2\rho_{x'y'}\sigma_{mo}\sigma_{so} + \sigma_{so}^2}}{\sigma_{mo}^2(1-\rho_{x'y'}^2) + 4\rho_{x'y'}\sigma_{mo}\sigma_{so}} f_R(r) \qquad (6.30)$$

Where

$$\sigma_{mo}^2 = \frac{\sigma_m^2}{(2\pi f_m)^2} \qquad \sigma_{so}^2 = \frac{\sigma_s^2}{(2\pi f_m)^2} \qquad (6.31)$$

The correlation coefficient $\rho_{x'y'}$ typically varies over a range of 0.5 to 0.9 for data taken by Loo.

The average fade duration is found from the level crossing rate from

$$AFD = \frac{1}{LCR_N} \int_0^R f_R(r)dr \qquad (6.32)$$

6.6   SUMMARY

In this chapter, an overview of the land mobile satellite system channel is given.   As a signal propagates from a satellite to a mobile receiver, it undergoes a variety of changes beyond the simple, free-space propagation loss.   Interaction with various layers and components of the atmosphere cause a number of effects, most of which are frequency dependent.   In the case of LMSS channel, the most serious signal degradations are caused by the signal interacting with objects near the vehicle, resulting in signal fading. In the majority of cases three types of fading will occur.   Rayleigh fading is the result of the vehicle receiving a large number of non-coherent reflected signal components caused by scatterers in the vicinity of the mobile receiver. In addition to these scattered components the LMSS channel will typically have some line-of-sight visibility to the satellite.   The addition of a direct signal component to the Rayleigh component creates a Rician fading channel.   As the vehicle travels trees often obscure this line-of-sight path. The foliage of the trees interacts with the signal, causing a Lognormal fade. Each of these signal components must be accounted for when modeling the LMSS channel.   Models have been proposed by several authors .   The Loo channel model (1985)was summarized in this chapter, because the steps in its development are applicable to other proposed models and because it has shown reasonable agreement to measured data.   Space limitations do not allow for an equivalent treatment of the other models.   Models by Lutz et al., Lutz (1986,1991), Smith and Stutzman (1986) should also be considered when modeling the LMSS channel.

REFERENCES

BOOKS
Beckmann P. and Spizzichino A., ,"The Scattering of Electromagnetic Waves From Rough Surfaces", Pergamon, Emsford, N, 1963.

Petr Beckmann,"Probability In Communication Engineering", Harcourt, Brace and World, 1967.

Flock W. L., "Propagation Effects On Satellite Systems At Frequencies Below 10 GHz, A Handbook For Satellite Systems Design" 2nd Edition, NASA Reference Publication
    1102(02),
1987.

JAKES W. C. (ED), "Microwave Mobile Communications", Wiley, New York, 1974.
W. C. Y. Lee, Mobile Communications Engineering, New York, McGraw-Hill, 1982.

A. Whalen, "Detection Of Signals In Noise", Academic Press, Inc., New York, NY 1971.

REPORTS
Campbell R. L. and Estus R., "Attenuated Direct And Scattered Wave Propagation On Simulated Land Mobile Satellite Service Paths In The Presence of Trees," Proceedings. Mobile Satellite Conference, JPL Pub. 88-9, Pp. 101 – 106, Pasadena, CA, May, 1988.

Clarke R. H., "A Statistical Theory Of Mobile-Radio Reception," The Bell System Technical Journal, July-August 1968, Pp. 957-1000.

Smith W. T. and Stutzman W. L., "Statistical Modeling For Land Mobile Satellite Communications", Virginia Tech Report EE Satcom 86-3, Virginia Tech, Blacksburg, VA, August, 1986.

Vogel W. J. and. Smith E. K, "Propagation Considerations In Land-Mobile Satellite Transmissions," MSAT-X Report 105, NASA-JPL, Pasadena, Ca, Jan. 1985

JOURNAL/SYMPOSIA ARTICLES
Amoroso F., and Jones W.W., "Modeling Direct Sequence Pseudonoise (DSPN) Signaling With Directional Antennas In The Dense Scatterer Mobile Environment" ,38th IEEE Vehicular Technology Conference, Philadelphia, PA, pp. 419 – 426, June, 1988.

Loo C., "A Statistical Model For A Land Mobile Satellite Link, IEEE Trans. Vehicular Technology". Vol. Vt-34, No. 3, August, 1985, pp. 122 – 127.

Loo C. and Secord N., "Computer Models For Fading Channels With Applications to Digital Transmission", IEEE Trans. Vehicular Technology. Vol. 40, No. 4, Nov. 1991, pp. 700 – 707.

Lutz E., Cygan D.,.Dippold M., Delainsky F. and Papke W., "The Land Mobile Satellite Communication Channel – Recording, Statistics, and Channel Model, IEEE Transactions Vehicular Technology". Vol. 40, No. 2, May 1991, pp. 375 – 386.

Rice S. O.," Mathematical Analysis of Random Noise, Part I", BSTJ Vol. 23, pp. 282 – 332, July 1944.

Rice S. O., "Mathematical Analysis of Random Noise, Part II", BSTJ Vol. 24, pp. 46 – 156, January 1945.

Rice S. O., "Mathematical Analysis of a Sine Wave Plus Random Noise", BSTJ Vol. 27, pp. 109 – 117, January 1948.

Vishakantaiah P. and Vogel W. J., "LMSS Drive Simulator For Multipath Propagation," Proceedings Of Napex XIII, JPL 89-26, San Jose, Ca, June, 1989, pp. 42-47.

Vogel W. J. and Hong U. S., "Measurements And Modeling of Land Mobile Satellite Propagation at UHF And L-Band," IEEE Trans. Antennas and Propagation, Vol. 36, pp. 707 – 719, May 1988.

Vogel W. J. and Goldhirsh J., "Fade Measurements at L-Band and UHF in Mountainous Terrain For Land Mobile Satellite Systems," IEEE Trans. Antennas and Propagation, Vol. 36, pp. 104 – 113, Jan. 1988.

Chapter 7

Payload Trades, Antennas and Communications

Robert A. Peters

Stellar Solutions, Inc

7.1   INTRODUCTION

A satellite is composed of two major subsystems, the payload and the bus. This chapter describes both subsystems with emphasis on the payload because it impacts the quality and the type of services that the satellite can deliver. The communications payload generates the revenue for the satellite operator. As a result, the operator of the satellite usually specifies the payload in detail while the builder of the satellite adopts the specified payload to an existing bus structure. The bus provides the electrical power, thermal stability, pointing (keeping the antennas pointed in the correct direction), station keeping (to keep the satellite from drifting out of its assigned position) and structural integrity. This chapter discusses the major topics associated with payloads used by GMSS satellites and some major bus subsystems.

7.2   BASIC FEATURES OF A PAYLOAD

The simplest type of payload is transparent or "bent pipe." A simple configuration of a three transponder payload is shown in Figure 7.1. A transponder covers a frequency band and has its own filters and usually its own amplifier.

The capacity of a satellite is frequently expressed in terms of the number of transponders, but since this does not determine either the bandwidth or RF power, it is not a good indicator of capacity. Indeed, there is no good figure of merit for a communications satellite. At least, in part, this is because regulatory bodies determine the bandwidth (which has no direct economic price) while engineers determine the RF power (which is the major determinant of the cost of a satellite).

With reference to Figure 7.1.

1. The signals are received at the Rx (receive) antenna.
2. A low noise amplifier (LNA) amplifies the signals.
3. The receiver amplifies and translates the signal frequency (usually from the uplink to the downlink frequency).
4. The signals are separated into frequency bands A, B, and C at the input multiplexer.
5. The signal is amplified again, from a few tenths of a milliwatt up to hundreds of watts through a high power amplifier (HPA).
6. The HPA outputs are filtered and combined at the output multiplexer.
7. The amplified signal is transmitted back to earth by the Tx (transmit) antenna.

Payloads have many possible variations. For example, the receiver can be after the input multiplexer, allowing different translation frequencies thereby providing more flexibility but requiring more hardware. A bent pipe payload receives, amplifies and routes the signal. The downlink signal is a replica of the uplink signal (except for some unavoidable additional noise and distortion) but translated in frequency. The frequency translation is necessary so that the downlink signal does not interfere with the much weaker uplink signal.

A processing payload demodulates the uplink signals. The payload can extract, modify, and route the signals as designed. Such systems use digital encoding where the analog signals are converted into a string of binary digits (bits). The "zeroes" and "ones" of the signals are recovered as "zeroes" and "ones." Information can arrive as a continuous stream or as data bursts or as packets. The packets are digitally switched and re-modulated as RF at the downlink frequency. Recovering the digital information at the satellite substantially improves the link performance, by reducing noise and distortion at the satellite; but more importantly, provides a substantial increase in flexibility. However, this flexibility in routing and control comes at the price of significantly increased system complexity and cost.

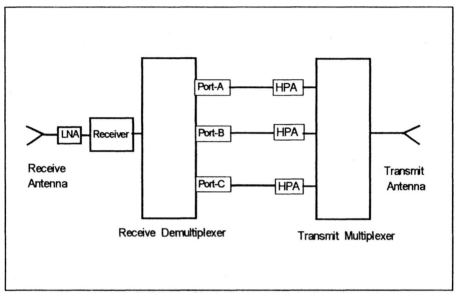

Figure 7.1 Block diagram of basic three transponder bent pipe payload

An additional dimension in flexibility can be provided by inter-satellite links (ISLs) communications network-in-the-sky. Where many satellites are connected through ISLs, a processing payload is essential to route the signal to its destination. IRIDIUM uses ISLs and digital switching to route a signal to any one of 48 beams on any one of 66 satellites.

Tables 7.1 and 7.2 compare the characteristics of several satellite systems in operation or under development, for voice communications to personal handsets. Table 7.3 provides a similar comparison for proposed high-speed multimedia data systems. Strictly speaking they are not GMSS but rather FSS (Fixed Satellite Services) systems, where the users are fixed (such as roof-top antennas). They are shown for completeness since the third generation (3G) GMSS are also moving into the high-speed data for mobile users.

Table 7.1. Some Personal Satellite Communications Systems with FCC license approval
(as of 1998)

	ACeS	Ellipso	Odyssey	Thuraya
Financial Partners	Pasifik Satelit Nusantara, Lockheed Martin+	Westinghouse, Harris, Israeli Aircraft Industries	TRW, Teleglobe	Thuraya Satellite Telecom of United Arab Emirates
Orbit (Type)	. GEO	Elliptical and equatorial	6471 kms circular	GEO
Spectrum	L & C	UHF	L, S, & Ka	L & C
Data rate	9.6 Kbps	0.3-9.6 Kbps	9.6 Kbps	9.6 Kbps
Beams/satellite	119-190		61	250-300
Start of operations	2000	-	-	2001
Number of [1] satellites	2	17	15	2
Status (as of Jan, 2001)	Operational	Proposed	Discontinued	Operational

[1] includes in-orbit spares but not ground spares

Table 7.2 Some Personal Satellite Communications Systems with FCC license approval

	Globalstar	ICO[1]	IRIDIUM	Orbcomm
Financial Partners	Loral, Qualcomm, Alcatel,	Teledesic, Craig McCaw, Hughes, NEC	Motorola Raytheon	Orbital Sciences, Teleglobe
Orbit (Kms)	6471	6459	484	484
Spectrum	L, S, & C	S & C	L and Ka	VHF
Data rate	7.2 Kbps	4.8-144 Kbps	2.4 kbps	56.7 Kbps
Beams/satellite	16	163	48	
Start of operations	2000	2003	1998	1995
Number of satellites	48	10	66	28
Status (as of Jan, 2001)	Operational	Under construction	Operational	Operational

[1] Data is for the "new" ICO

Table 7.3. Proposed Multimedia Broadband Data Systems (FSS) (as of 1998)

	Astrolink	**Skybridge**	**Spaceway**	**Teledesic**
Financial Partners	Lockheed	Alcatel/Loral	GM-Hughes	Bill Gates, Craig McCaw, Boeing
Application	Data, video, rural telephony	Voice, data, video-conf.	Data, multimedia	Voice, data, video-conf.
Orbit (Km)	GEO	911	GEO	LEO TBD
Spectrum	Ka	Ku	Ka	Ka
User Antenna (inches)	33-47		26 and larger	10
Data Rate	Up to 6 Mbps	16 Kbps to 2 Mbps up, 16 Kbps to 60 Mbps down	Up to 6 Mbps	16 Kbps to 64 Mbps
System	$4 B	$3.5 B	$3.5 B	$ 9 B
Number of Satellites	9	64	8 initially	288
Access method	FDMA TDMA	CDMA, TDMA, FDMA, WDMA	FDMA, TDMA	MF-TDMA, ATDM
ISLs	yes	no	yes	Yes

7.3 ANTENNAS

The most important parameter of a spacecraft antenna is its gain, which is a measure of how well an antenna focuses outgoing power or collects incident power. A larger (higher gain) antenna transmits a narrower beam and collects/receives a higher level of incident power than a smaller antenna.

Antenna gain, G, is defined by

$$G = \frac{4\pi\eta A}{\lambda^2} \qquad (7.1)$$

Here η is the antenna efficiency, which typically is 60 to 70%, A is the antenna aperture area, and λ is the RF wavelength. Receive and transmit antenna gains are denoted as G_R and G_T respectively.

A high gain antenna produces a greater flux density on the ground for the same RF power, but over a smaller region (conservation of energy). It will detect a weaker uplink signal from the ground, but over a smaller coverage region than a lower gain antenna.

The antenna gain is the ratio of the illuminated area to the area of a sphere of the same radius. This is the same as the ratio of the power of the coverage area to the power in the coverage area if the energy were radiated evenly in all directions (isotropically). This is illustrated in Figure 7.2 and Equation 7.2. The antenna gain, in dB, can be calculated by knowing the antenna diameter in meters, the radiated (or received) frequency in MHz (f_MHz)and the antenna efficiency. Antenna gain is measured in units of dBi in order to make the comparison to an isotropic antenna explicit. Equation 7.2 is obtained using $\lambda=c/f$, $A=\pi d^2/4$ where d is the antenna diameter in meters and c is the speed of light in m/sec and placing in logarithmic format.

$$G(dBi) = 20\log(d) + 20\log(f\_MHz) + 10\,Log(\eta) + 20\log(\pi/c) \quad (7.2)$$

$$= 20\log(d) + 20\log(f\_MHz) + 10\,Log(\eta) - 39.6$$

The dependence of transmit antenna gain on wavelength follows because the smaller the wavelength, the narrower the beam. Conservation of energy requires that the energy collected by the receive antenna and hence antenna gain depends only on its area and not on the wavelength, in apparent disagreement with Equation 7.1. The explanation will be given in the following subsection. A very useful approximation for the beamwidth, θ, defined as the angle where the gain of the antenna is half that of its peak value is:

$$\theta(degrees) \approx 70\,(\lambda/d) \quad (7.3)$$

where λ is the RF wavelength and d is the diameter of the antenna.
If the satellite antenna can discriminate between two polarizations, the number of transmit and receive beams within the bandwidth can be doubled, making better use of the limited spectrum. The ability of the antenna to discriminate between two signals of the same frequency and in the same beam, but of different polarization is called the cross-polar isolation. Co-polarization isolation is another important antenna specification when the payload has different beams reusing the same frequency.

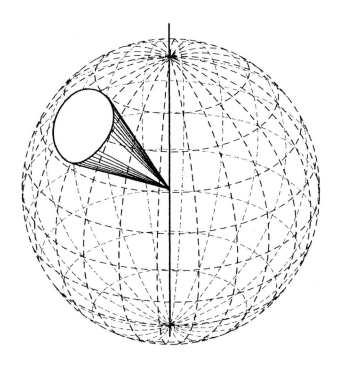

Figure 7.2 . A point source inside the sphere will illuminate the entire area of the sphere, $4\pi r^2$. If the radiation is focused only a small fraction of the area (shown in white) is illuminated. The antenna gain is the ratio of the illuminated area to the entire area

The desired shape for a beam is rarely a single circular pattern on the ground. Usually the desired shape follows the contours of the landmass or of the geographic boundaries as shown in Figure 7.3. The desired shape can be obtained by using multiple small beams. Rather than combining the circular beams into a composite beam, they can be left as independent beams. Then the RF energy of a single signal is not spread out over the entire coverage area but concentrated over a narrower area and the antenna gain can be much higher than a single, circular beam covering the same area. Table 7.4 summarizes the most important parameters in specifying an antenna.

The simplest form of a payload with multiple beams is shown in Figure 7.4. A three-transponder payload uses one transponder per coverage circle. There is no connectivity between coverage areas in this simple payload. The payload shown in Figure 7.4 could be designed so each transponder illuminate three coverage circles and provide connectivity; but each transponder would cover three times the area with one-third the gain.

Table 7.4. Key Antenna Performance Specifications

Gain (G)	Ability of antenna to focus a beam, see Equation 7.1.
G/T	Antenna gain divided by the noise temperature. Determines receive sensitivity.
Cross-polarization isolation	Ability of an antenna to discriminate between orthogonal polarizations.
Co-polarization isolation	Ability of an antenna to separate the signals from different beams.
Gain flatness	Uniformity of the antenna gain is over the specified coverage region.

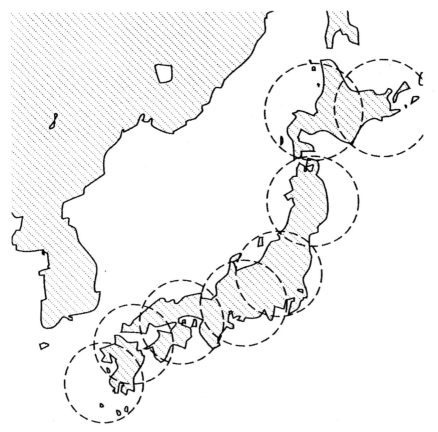

Figure 7.3. Illustration of shaping antenna coverage using multiple circular beams. The coverage can be a single composite, shaped bream.

This relatively simple change to multiple small beams has significant consequences.

1) The area covered by each beam is much smaller, increasing the antenna gain and allowing smaller, less expensive ground terminals.
2) The same total RF power can be used to carry more traffic, and/or reduce the RF power.
3) The same bandwidth can be used for multiple beams, greatly increasing the available bandwidth of the satellite.
   a) This allows the same bandwidth to be reused, increasing the amount of traffic that can be accommodated within the bandwidth.
   b) A terminal will have to be tuned to the correct frequency to function and re-tuned if it is moved.
4) Connectivity between beams, if required for the mission, must be provided by additional hardware on the satellite since a single uplink does not encompass the entire coverage area.

The beams may be formed by individual feeds as shown in Figure 7.3 or by a mechanical or electronic beam former. Mechanical beam formers use fixed wave-guide components to control the RF phase and amplitude. Usually there is one amplifier for each transponder in each composite beam. Electronic beam formers use electronic RF phase shifters (and sometimes electronic amplitude control) to produce the multiple beams. There are many radiating elements and each element usually has its own amplifier.

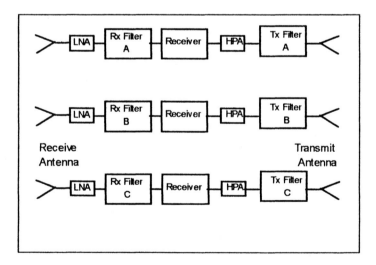

Figure 7.4 Illustration of a Multiple Beam Coverage Transponder

Thus, there are large numbers of amplifiers in such a beam former. Each amplifier has to work over the entire transmitted bandwidth and carry multiple signals destined for various beams. This requires a highly linear operation, which results in a lower efficiency system. The flexibility allowed by electronic beam formers can offset the low efficiency. Energy can be shifted from beams that are lightly loaded to beams that are heavily loaded, increasing the flexibility of handling varying traffic loads. Beam size can be enlarged, reduced, or the beams can be moved dynamically.

Advances in software have made it possible to have a shaped coverage without a beam-forming network. Shaped antennas have lower RF losses, reduced cost and reduced mass compared to beam forming antennas. The desired coverage is achieved by shaping the reflector so that the antenna coverage has the desired shape. This type of antenna, not surprisingly, is called a shaped reflector antenna. The shaped beam technique does not allow for frequency reuse in a coverage area.

Antennas fall into two categories, reflector antennas and direct radiating antennas. Within these two categories there are numerous subcategories. Reflector antennas use a feed that radiates the energy toward the reflector where it is then reflected toward the target. The size of the reflector is used in calculating the antenna gain. Direct radiating antennas radiate the energy directly toward the target. The size of the array is used to calculate the antenna gain. A large array is much more difficult to construct than a reflector so most satellites needing high gain antennas use reflectors.

The spot beam of an antenna can be pointed in various directions within a cone characterized by the scan angle. A direct radiating array can radiate at large scan angles as illustrated in Figure 7.5. Such arrays are attractive for low earth orbit (LEO) communication satellites because they operate over larger scan angles than reflector antennas. A geosynchronous satellite requires a scan angle of 17 degrees to cover the entire earth while a 780 km LEO requires a scan angle of 63 degrees (see Figure 7.5) to cover its field of view.

The increased scan angle of LEO systems comes at a cost. Since antenna gain is a function of the effective antenna diameter, the effective antenna gain is reduced when the scan angle increases as given in Equation 7.4.

$$\text{Gain } (\theta) = G \bullet [\cos (\theta)]^{1.3} \qquad\qquad (7.4)$$

In addition to the fact that antenna gain decreases as the scan angle increases, polarization purity also decreases. The additional cost, mass and power consumption of a direct radiating antenna make the reflector antenna the usual choice except for LEO satellites where the large scan angle requires a direct radiating antenna. No commercial GEO satellite currently uses direct radiating antennas. At MEO the situation is less clear and both direct radiating and reflector antennas have been proposed for this orbit.

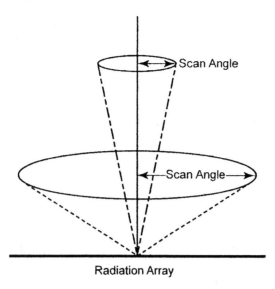

Radiation Array

Figure 7.5. Large and Small Antenna Scan Angles

7.4   LINK BUDGETS

A satellite link is a connection between a ground and a satellite antenna that allows information to be exchanged. The received energy per bit, E_b, must be sufficiently above the noise level, N, to allow the received signal to be recovered adequately. The important parameter is the ratio E_b/N. The noise is measured over the bandwidth used to transmit the bit. The link budget is used to calculate E_b/N. It is used to determine the antenna size and RF power needed for the link.

The two most important parameters that characterize RF performance of a spacecraft are the EIRP and the G/T. The EIRP is the Effective Isotropic Radiated Power and is the product of the transmitting antenna gain, G_T, and the transmitted RF power, P_T, or:

$$EIRP = G_T P_T. \tag{7.5}$$

The EIRP is proportional to the power density leaving the spacecraft. The link noise is characterized by a "system noise temperature," Tsys, whose units are degrees Kelvin, or °K. Thus the figure of merit for the receive antenna is:

$$G_R/T = \text{(receive antenna gain)}/(\text{Tsys}) \tag{7.6}$$

Reducing the system noise power level by half is equivalent to doubling the antenna aperture area . For a spacecraft antenna focused on the earth, the noise temperature has to be greater than the earth's temperature. The earth station antenna looking into space sees a much lower temperature. Rain, in addition to attenuating the spacecraft's signal, increases the thermal noise level seen by the earth station. Hence E_b decreases and N increases during a rain storm.

Receive antennas have a special problem when the sun comes into their field of view. The noise temperature increases and can bring the link margin below acceptable levels ('breaks the link'). Some antenna operators point the antenna away from the sun (and the satellite) to keep the antenna from concentrating heat on the antenna feed and damaging it.

Returning to link budget calculations, the path loss, L, between the transmit and receive antennas, is just the spread of the energy ($1/4\pi R^2$) loss with an additional term proportional to λ^2.

$$L = \left(\frac{\lambda}{4\pi R} \right)^2 \qquad (7.7)$$

Conservation of energy requires that the flux density depend only on distance and not on wavelength as in Equation 7.7. The reason for the wavelength dependence in the equation term becomes clear when the full expression for the link budget is examined.

$$P_R = (EIRP) * L * G_R \qquad (7.8)$$

P_R is the received power (also called S, the signal strength) and G_R is the gain of the receiving antenna. The received power is the transmit effective power (EIRP) multiplied by the path loss factor L (L<1), multiplied by the receive antenna gain. It is usually expressed using logarithms.

$$P_R = 10\log(P_T) + 10\log(G_T) - 20\log(4\pi) - 20\log(R) + 20*\log(\lambda) + 10\log(G_R) \qquad (7.9)$$

We previously noted that there was frequency dependence in G_R and L that did not seem reasonable. The resolution lies in examining the product of $L* G_R$ (where G_R is given from Equation 7.1).

$$L * G_R = \left(\frac{\lambda}{4\pi R} \right)^2 \frac{4\pi\eta A}{\lambda^2} = \frac{\eta A}{4\pi R^2} \qquad (7.10)$$

The product of L* G_R is independent of wavelength. The wavelength dependence is put in the definition of receive antenna gain and in the loss term so that they cancel each other out. The advantage of this method is that it allows a single definition of gain for both transmit and receive antennas.

Returning to the link equation, we note that Equation 7.9 is incomplete in that it gives the signal strength but not the noise. We are looking for an expression for Eb/N.

To convert Equation 7.6 to the same units of energy as Equation 7.9, Tsys must be multiplied by the Boltzmann constant "K" = 1.38 x 10^{-23} Joules/Degrees-Kelvin. Converting K into a logarithmic form, Equation 7.9 becomes, after equating No with the thermal noise component of N:

$$P_R/No = S/No = 10\log(P_T) + 10\log G_T - 20\log(4\pi) - 20\log(R) + 20\log(\lambda)$$
$$+10\log(G_R) - 10\log(Tsys) + 228.6 \qquad (7.11)$$

Dividing the power by *the data rate (R_b)* in bits per second, gives the energy per bit Eb = P_R/R_b.

$$Eb/No = (P_R / R_b)/No = 10\log(P_T) + 10\log G_T - 20\log(4\pi)-20\log(R)+$$
$$20\log(\lambda) + 10\log(G_R) - 10\log(Tsys) - 10\log(R_b) + 228.6 \qquad (7.12)$$

Eb/No specifies the link quality in a way that is independent of the bit rate and can directly be compared to the modem requirements. Another combination of terms that is used frequently, in regulatory areas is the PFD or Power Flux Density. The PFD is used to specify how much power one system can radiate into other terrestrial and space systems. There is no wavelength dependence of the PFD. It is given by:

$$PFD = \frac{EIRP}{4\pi R^2} \qquad (7.13)$$

Links must be designed to cope with some degradation and fade. The additional power over the minimum necessary to maintain the link is called the link margin. Other degradations include interference from other space and terrestrial systems and antenna co-polar and cross-polar interference. Rain attenuation is different from other degradations because at the ground station the rain attenuates antenna co-polarization interference as well as interference from nearby spacecraft.

Of the various factors that can degrade a link, rain is the most important, especially at frequencies above 6 GHz. Rain attenuation or rain fade is given statistically as the percentage of time a given rain fade will be exceeded. The rain fade statistics for the areas of the world have been collected for various RF frequencies. Rain fade generally increases as the RF frequency increases, but the increase is not monotonic. There are discrete frequencies where water absorbs RF energy at high rates. The requirements are usually stated as a percentage of availability, such as a 99.94% availability for a given rain region. The greater the link margin the higher the availability.

A typical link budget is shown in Table 7.5 from an FCC filing.[1] Here Io is the interference contribution and is added to the thermal noise No to get the total noise. The PFD is given as a spectral density over 4 kHz, per regulatory requirements.

7.5   POLARIZATION AND FREQUENCY REUSE

Using two polarizations allows the same RF bandwidth to be used twice. Polarization can be linear or circular. Linear polarization can be either vertical or horizontal and a circular polarization is either left hand or right hand polarized. The question of which type of polarization is "better" is the subject of endless debates.   Circular polarization does not require any alignment of the antenna by the customer, which is a major advantage for handheld and consumer applications, but has a somewhat greater cost

LEO satellites have poor polarization purity due to their large scan angles and so cannot take advantage of polarization reuse. If a LEO satellite is constructed with many antennas, each covering a small area, then it could obtain good polarization isolation and take advantage of polarization reuse. However, this solution may not be cost-attractive.

---

[1] "Application of Iridium LLC to Launch and Operate the MACROCELL Satellite system" to Federal Communications Commission, 26 September 1997.

Table 7.5 Typical Link Budget for a LEO Satellite

FDMA/TDMA				
	Nadir Beam		Outer Beam	
	Uplink	Downlink	Uplink	Downlink
Frequency (MHz)	2007.5	2182.5	2007.5	2182.5
Elevation (degrees)	90.0	90.0	15.0	15.0
Transmit Antenna Directivity (dBi)	0.0	19.8	0.0	27.8
EIRP (dBW)	4.0	21.6	4.0	29.6
Free Space Loss (dB)	157.1	157.8	165.1	165.8
Atmospheric / Polarization . Loss (dB)	0.8	0.8	0.8	0.8
Shadowing Loss (dB)	15.0	15.0	15.0	15.0
Total Path Loss	172.9	173.6	180.9	181.6
Receive Antenna Directivity (dBi)	19.8	0.0	27.8	0.0
Tsys (dB K)	27.7	24.8	27.7	24.8
G/T (dB/K)	-7.9	-24.8	0.1	-24.8
Data Rate (Kbps)	34.5	34.5	34.5	34.5
Eb/No Thermal (dB)	6.4	7.4	7.4	7.4
Eb/Io	18.0	18.0	18.0	18.0
Eb/(No+Io)	6.1	6.1	6.1	6.1
Rqd Eb/(No+Io) (dB)[1]	6.1	6.1	6.1	6.1
Margin (dB)	0.0	0.0	0.0	0.0
PFD (dBW/m^2/4 kHz)	NA	-114.4	NA	-114.3
(1) includes modem implementation loss				

7.6 ACCESS TECHNIQUES: FDMA, TDMA, CDMA

How can several hundred different signals share the same amplifier and yet remain separate? This is made possible by the several different access techniques. There are three distinct access methods used today and many more combinations of these methods.

FDMA, or frequency division multiple access, uses a separate frequency assignment for each signal. Several signals, each at a different frequency share the amplifier. Radio and TV are the best known illustrations of this method. A three-carrier FDMA case is illustrated in Figure 7.6. The bandwidths need not be uniform and in practice are sized according to the required capacity. The most common form of FDMA is SCPC or single carrier per channel.

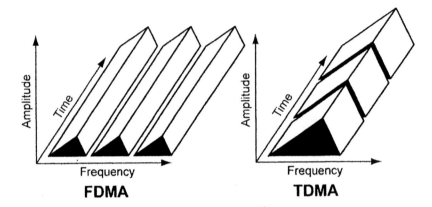

Figure 7.6 Illustration of Multiple Access Using FDMA and TDMA Techniques

TDMA, or time division multiple access, allocates the entire amplifier to a specific user for a certain amount of time, called a burst. Figure 7.6 illustrates a three-burst TDMA frame. [For the same composite data rate total TDMA bandwidth used will be the same as for the FDMA example]. Multiple users share the amplifier in time rather than by frequency as in FDMA. The burst pattern is repeated regularly. One complete cycle is called a frame. Users requiring more capacity are given longer bursts than users requiring less capacity. IRIDIUM employs an interesting variation. Uplink bursts alternate with downlink bursts, both on the same frequency. The uplink and downlink are separated in time but not frequency.

CDMA or code division multiple access is the newest technique and the most difficult to explain. Several signals share the same frequency at the same time but each signal is sent with a unique code. All signals are received together and not separated in time or frequency. The combined signal looks like noise but when the aggregate of signals is multiplied by the right code its signal is recovered. Imagine the digital signal is composed of −1 and 1 rather than 0 and 1. For this example, we will assume a 4 bit code. Four possible codes are illustrated in Table 7.6. For purposes of this discussion we label the codes A, B, C and D.

Table 7.6 Four Possible CDMA Codes

	Code
A	1111
B	1-11-1
C	11-1-1
D	-111-1

A customer using code A, consisting of 1111 (shown in Table 7.6) sends a "1" by multiplying 1 by code A or:

$$1 \times (1, 1, 1, 1) = 1, 1, 1, 1$$

as illustrated in Table 7.6. A "0" is always treated as a -1, and transmitted as

$$-1 \times (1, 1, 1, 1) = -1, -1, -1, -1$$

A different customer is transmitting using a different code, say code B (1-11-1 as given in Table 7.6). A "1," is transmitted as 1, -1, 1, -1 and a "0" as -1, 1, -1, 1.

The customer using code A will receive his signal and the signal of customer B (and any other customers). All signals will be multiplied by code A and the four bits added together. The customer using code A will receive a 1 as a 4 and a -1 as a -4. Subscriber B, C and D's signals will sum to 0 when multiplied by code A as shown in Table 7.7. All non-desired signals cancel (ideally). Each user recovers only their signal and the four users in this example can share the same frequency at the same time. Four bits are transmitted for each desired bit using the same bandwidth as FDMA. The Globalstar system uses CDMA and it is widely used for terrestrial cellular phone service.

The non-desired signals add some undesired noise since the cancellation will not sum exactly to zero. To keep the total noise down, power control must be implemented. The poorer the power control, the fewer users the CDMA system can accommodate. This is a major problem for satellite links where the propagation delay limits the power control accuracy in a dynamically changing link.

Table 7.7. Use of CDMA Codes

Informa-tion bit	Code	Sent bits	Receive code	Sent Bits x code	Sum of bits
1	A (1111)	1,1,1,1	A(1111)	1,1,1,1	4
-1	A (1111)	-1,-1,-1,-1	A(1111)	1,1,1,1	-4
1	A	1,1,1,1	B(1-11-1)	1,-1,1,-1	0
1	A	1,1,1,1	C(11-1-1)	1,1,-1,-1	0
1	A	1,1,1,1	D(-111-1)	-1,1,1,-1	0
1	B(1-11-1)	1,-1,1,-1	B(1-11-1)	1,1,1,1	4
-1	B(1-11-1	-1,1,-1,1	B(1-11-1)	-1,-1,-1,-1	-4
1	B	1,-1,1,-1	A(1111)	1,-1,1,-1	0
1	B	1,-1,1,-1	C(11-1-1)	1,-1,-1,1	0
1	B	1,-1,1,-1	D(-111-1)	-1,-1,1,1	0

There are various combinations of access techniques. For example, a single FDMA signal can itself be a TDMA signal. In this case there can be several TDMA signals in one transponder. There can be several FDMA carriers, each with multiple CDMA signals. In TD-CDMA, or Time Division CDMA, several CDMA signals are transmitted in each burst. A detailed discussion of these hybrids and trade-offs is beyond the scope of this book and is the subject of as much emotional as technical evaluation.

Table 7.8 summarizes the author's opinion of the strengths and weaknesses of each access approach. CDMA has become popular for terrestrial cellular, in part because of its higher capacity in situations where the traffic distribution per base station is fairly uniform. However some of its disadvantages, particularly power control and limited peak beam capacity, are more severe for satellite application than for terrestrial cellular.

7.7 CODING: TRADING BANDWIDTH FOR POWER

Coding allows additional bandwidth to be used to reduce the RF power needed to obtain a desired link margin. Coding transmits additional bits that allow errors to be detected and corrected. The simplest type of code is hard decision block coding. The following example gives an intuitive idea of how errors can be corrected by coding. The nine digits (the information bits) sent are shown in the unshaded boxes in Table 7.9.

Table 7.8. FDMA/TDMA/CDMA Trade-offs

	FDMA	TDMA	CDMA
Ground Station Complexity	Simplest	Complex	Complex
Satellite Amplifier Linearity	Very high	Low	Low
Ground HPA power required	Low	Highest	Low
Required Link power control	Low	Low	Stringent
Resistance to Interference	Poor	Good	Good
Soft handovers (non-GEO)	Difficult	Difficult	Easy
Bandwidth on Demand	Difficult	Moderate-difficult	Easy
Ease of Regeneration	Moderate	Easiest	Hardest
Analogue Signals Possible?	Yes	No	No
Peak beam capacity	Higher	Higher	Lowest
Average Capacity	Lower	Lower	Highest

For each row and column, parity bits are sent as shown in the Table. A parity bit is 0 if the sum of the row or column bits is even and 1 if the sum is odd. For nine bits of information, 15 bits must be sent.

Table 7.9. Example of Hard Decision Block Coding, No Errors

1	0	1	0
1	1	1	1
0	0	1	1
0	1	1	

Information bits are shown unshaded. Error correction bits are parity bits and are shown lightly shaded.

Table 7.10. Example of Hard Decision Block Coding, with one Error

1	**0**	1	0
1	**0**	1	1
0	**0**	1	1
0	1	1	

Error bit shown in Bold. Error correction parity bits are lightly shaded. Darker shade row and column will have parity error

The error correction circuitry determines that there is an error by calculating the row and column parity bits of the received 9 bits and comparing them to the transmitted parity bits (shaded row and column in Table 7.10). The specific bit in error can be found by the intersection of the row and columns containing parity errors (the lightly shaded rows and columns in Table 7.10). What happens if the parity bit is an error rather than in the data bit? Then there would be an error in **either** the row or the column but not both. The code discussed here is not used because it is inefficient.

Too many error-correcting bits are required. It was introduced to provide an intuitive explanation of how coding works.

The most powerful error correcting codes are soft decision convolution codes. Soft decision codes do not classify a bit as a "0" or a "1," but as a number with several possible values, most commonly a three bit number (0 to 7) that represents how close the number is to a "0" or "1." A "0" means that there is little noise present so the modem is "very sure" the number is a "0." A "3" or "4" would mean that a lot of noise was present and the modem is not sure if the bit was a "0" or a "1." A "7" would mean that the modem is very sure that the number is a "1." The knowledge of which bit is more likely to be in error allows much more efficient error correction. These convolution codes gain much additional power by using statistical processing of nearby bits to determine the most likely value of a questionable bit.

A frequent source of confusion results from the two different bit rates that exist in coded signals. The information bit rate (the 9 bits of desired information) are the unshaded bits in Table 7.9 and Table 7.10 The transmission bits are the 15 shaded **and** the unshaded bits in the Tables. Eb always refers to the energy per information bit. The effect of coding on the required RF power can be obtained for Eb/No versus bit error rate (BER) curve for the particular code chosen. Coding is chosen to optimize system performance within the bandwidth constraints.

7.8   SPACE HIGH POWER AMPLIFIERS

High Power Amplifiers come in two varieties, travelling wave tube amplifiers (TWTAs) and solid state power amplifiers (SSPAs). TWTAs were the first microwave amplifiers to be used in communication satellites. They are high frequency vacuum tubes.

SSPAs, or solid state power amplifiers usually use GaAs FETs (field effect transistors). SSPAs were expected to be substantially more reliable than TWTAs, but early units in the 1970's proved less reliable than tubes. Reliability of both TWTAs and SSPAs has improved greatly since then and selection is now based on design requirements.

Table 7.11 Comparison of Advantages of SSPAs and TWTAs

SSPAs	TWTAs
Better efficiency for low power applications	Higher basic efficiency for moderate and high power applications
Lighter	Efficiency decreases less with frequency
Smaller	Efficiency decreases less with decreased drive and output power
Lower purchase price	

The mass and cost of TWTAs do not change dramatically with RF output power nor does frequency decrease very much with increasing frequency while the opposite is true for SSPAs. In general, TWTAs are favored for high frequency, high power applications and SSPAs are used for the lower power, lower frequency applications. Phased array antennas are usually made up of many low power transmitter modules and thus are usually powered by SSPAs.

7.9   TRANSPARENT VS. REGENERATIVE PAYLOADS

Most commercial satellites today are "bent pipe" or transparent satellites; they receive a signal, amplify it and retransmit it back to the ground. The retransmitted signal is the same as the receive signal, translated in frequency and with some unavoidable distortion added. The signal can be either analogue or digital.

A regenerative satellite recovers the actual bits (0 and 1). The signal is usually transmitted in packets where each packet has an address containing its destination. This address can be read on the spacecraft and the packet routed to the appropriate downlink or crosslink beam, substantially increasing system flexibility.

A Teledesic satellite (as first proposed) has 576 beams, or cells as Teledesic refers to them[2]. How can a ground station uplink signal be routed to the correct downlink? For simple systems with few beams each uplink bandwidth could be connected to a downlink beam. The selection of the uplink frequency would determine the downlink beam. For Teledesic this would require an uplink bandwidth 576 times the downlink bandwidth. This

[2] "Application of Teledesic Corporation for a Low Earth Orbit Satellite system in the Domestic and International Fixed Satellite Service" to Federal Communications Commission, Washington DC, dated March 21, 1994 and received March 24, 1994, page 48

could not work with the limited spectrum available. Imagine now that there are ISLs connecting the spacecraft. If there were 288 spacecraft, or over 147,000 uplink channels (288 x 576) would be required. Every MHz to the subscriber would require 1.47 GHz of bandwidth from the gateway station. Recovering the baseband signal allows electronics aboard the spacecraft to "read" the destination of the packet and route it appropriately. Satellites with many beams require baseband switching and hence regeneration.

A second major advantage of regeneration is link improvement. If link performance is equally uplink and downlink limited, regeneration can dramatically improve performance because the BERs of the up and downlinks add together. It removes the distortions and noise degradation from the uplink. Regeneration "cleans up" the signal unless the degradation is large enough to result in an error. Without regeneration the uplink degradations are added to the downlink degradations. The two degradations, neither of which is large enough to cause an error, could add together to result in an error.

Regeneration can relax the antenna co-pol and cross-pol requirements. This relaxation is especially important when there are many beams, as each co-frequency beams interferes with every other co-frequency beam. Regeneration solves both antenna and connectivity problems.

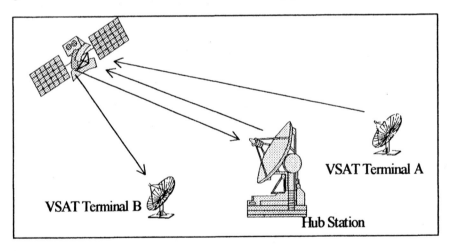

Figure 7.7. Illustration of VSAT link from Terminal A to Terminal B using a hub station.

Many small terminals, such as VSATs (very small aperture terminals) rely on having one near noiseless link so that almost all the link noise comes from the small terminal. One VSAT usually cannot talk directly to another VSAT through a non-regenerative satellite. There must be a VSAT - satellite - Gateway - satellite - VSAT link as shown in Figure 7.7 .

With regeneration, a VSAT to satellite to VSAT link becomes much easier. IRIDIUM, using regeneration, allows a phone to satellite to phone link without going through a gateway. Globalstar, which does not use regeneration, requires a Gateway connection to link the two handsets.

Table 7.12 On-Board Regeneration of Signals

Pros	Cons
Needed when there are high connectivity requirements.	Less flexible, can only work with predetermined bit rates and modulation and modulation formats
Improves link margin	Consumes substantial DC power
Higher interference rejection	Longer and more expensive design required due to ASICS and compatibility

7.10   INTER-SATELLITE LINKS

A Geosynchronous satellite can "see" about 40% of the earth's surface. Because each satellite sees so much of the earth, no commercial GSO satellite to date has an intersatellite link (ISL). ISLs for GSO satellites are being considered for network connectivity, particularly for data systems (Table 7.3).

If connectivity over a large geographical area is required, the smaller footprint of LEO satellites requires that they have a ground station in the coverage area of each satellite, a "store and forward" capacity, or intersatellite links (ISLs). Globalstar has a gateway under every satellite footprint in the service area. IRIDIUM, using ISLs, can offer service "anywhere, anytime". ISLs may use either RF or optical technology. Optical ISLs have very high antenna gain because of their short wavelength. Laser diodes have frequencies of the order of 1 to 6 x10^{15}, or about 10^5 higher than a 30 GHz RF wave. From Equation 7.2, the optical antenna gain is 100 dB more than a RF antenna of the same size and efficiency. A 10 cm. diameter optical antenna has 80 dB more gain than a one meter RF antenna. This allows a one-watt laser diode to provide the same flux density as a 100 Megawatt RF source! Optical detection is less efficient than RF detection so the above overstates the actual comparison but gives a clear idea of the advantages of an optical system. To appreciate the narrowness of the optical beam, a one-micron laser using a 10 cm. antenna would have a beamwidth of about 1 milliradian using Equation 7.3. Such a beam leaving Washington DC would be of the order of 10 meters when it reaches New York City. The very narrow beamwidth requires a closed loop tracking system, where each ISL "tracks" the signal from the communicating satellite

The narrow optical beam gives rise to several problems for optical crosslinks. These are:

- Difficult acquisition – the very narrow beam requires very precise knowledge of the location of the two satellites establishing the link and a special acquisition mode.
- Sensitivity to spacecraft vibration – A fast tracking system (of the order of 100 Hz) is necessary so that naturally occurring vibrations (from thruster valves, momentum wheels, solar array drives, etc.) do not degrade performance.
- Point ahead – The finite speed of light requires that the tracking system point to where the communicating satellite will be when its signal reaches it, not where it was when it received the signal. This is particularly true for GSO satellites.
- Optical cleanliness – The antenna gain is so high that even small amounts of contamination can significantly degrade it. The optical system

must be kept clean through ground test and launch and from thrusters used for station keeping.

Solutions for these problems have been demonstrated and are not considered "show stoppers", however optical ISLs do require extra care and some additional steps during spacecraft integration and test. Optical ISLs are well suited to applications that require high data rates over large distances. These include communication between widely spaced satellites, particularly geosynchronous satellites and communication links with high data rates. This is because, unlike RF links, an optical ISL's mass and power increases only slightly as the data rate and separation distance increases. The optical tracking system requirements do not change as capacity increases since the optical power is only a small fraction of the total ISL power requirements.

A requirement for an optical or RF ISL to operate through solar conjunction (when the sun is within the field of view of the ISL antenna) is a major ISL design driver. The sun energy increases the noise level when it comes within the field of view of an optical or RF antenna. Operation through solar conjunction is possible with both microwave and optical systems but is actually easier with an optical system.

Table 7.13. Comparison of Microwave and Optical ISLs

Microwave ISLs	Optical ISLs
Requires less mass and power for low capacity, short links	Less mass and power for high capacity, long distance links
Interference coordination with other systems is difficult	Interference with other system unlikely, easier to control
Difficulty of working through solar conjunction increases with frequency	Can work through solar conjunction with proper filtering
Slow simple tracking, or even open loop tracking possible	High speed tracking with point-ahead correction required

7.11   THE SATELLITE BUS

To a bus engineer the payload is a large electrical heater in the spacecraft that must be maintained under benign conditions in the harsh environment of space. Despite the external temperature extremes, the payload must be kept close to room temperature and be continuously supplied with power regardless of its availability from solar cells.

The bus subsystems, which are discussed in this chapter, are:
- Structure
- Power
- Thermal
- Tracking, telemetry, control, and ranging
- Altitude control and propulsion

7.11.1   Structure

The structure is the "house" of the satellite. Most commercial satellites are deployable with solar array and antennas folded to take up less room on the launch vehicle and unfolded (deployed) in space.

There are two basic types of structures: a spin stabilized satellite (spinner) and a body stabilized spacecraft. The early communication satellites were predominantly spinners. Spinners are cylindrical and usually have a "despun" inner platform, which spins at exactly one revolution per day so it is always pointed toward the same point on the earth. The outer shell, which contains the solar cells, spins in the opposite direction as the despun platform. A spinning platform makes attitude control easy (the spacecraft is one large gyroscope). It provides a uniform temperature since the outer shell is rotating between temperature extremes. The curvature of the solar array decreases the electrical power from the solar cells by a factor of π over what it would be if the array was flat and pointing toward the sun. Increasing the solar array area requires awkward telescopic extensions of the satellite.

By contrast, a body-stabilized satellite is square or rectangular in shape. It maintains a position that is constant relative to the earth by rolling once an orbit. Its antennas are always pointed toward the earth and its flat solar array "wings" rotate so that they are always pointing toward the sun. If additional power is needed, more panels can be added to the array. All satellites for mobile communications are body stabilized in large part because of the high power requirements.

7.11.2   Power Subsystems

Commercial satellites use solar cells to generate electrical power. Over the life of the satellite, the power output of the solar cells degrades due primarily to proton radiation from the sun and electrostatic discharges. The satellite must meet its power requirements at the end of life (EOL) thus requiring solar arrays that are larger than needed at the beginning of life (BOL). Unfortunately, most satellites have their peak demand for power at EOL when the satellite is fully loaded. GaAs solar cells are more efficient and degrade less than Si solar cells, but cost substantially more to procure. They have only recently been used in commercial satellites.

Whether in LEO or GSO, a satellite will usually spend some of its time in the shadow of the earth (eclipse). During an eclipse the power must be met entirely by batteries that have to be recharged when the spacecraft is in sunlight. During the early 1980's nickel hydrogen batteries replaced nickel cadmium batteries for spacecraft. The eclipse for a GSO satellite can last up to 72 minutes but never occurs more than once a day. LEO satellites have shorter but more frequent eclipses. They have less time to charge their batteries between eclipses and go through more frequent charge and discharge cycles. This leads to substantial differences in the power system designs between LEO and GSO satellites. A LEO system can also use battery power collected over a low coverage area such as an ocean, to augment the solar array power to meet short peak demands.

7.11.3   Thermal Subsystem

The thermal subsystem must dissipate the heat generated primarily by the payload but also other bus components. Heat pipes distribute the heat evenly over the north and south surfaces of the satellite where it is radiated into space. These surfaces never directly face the sun. Optical surface radiators (OSRs) mounted on these surfaces are shielded from most of the sun's radiation. The payloads on some satellite designs are limited by the amount of heat that can be radiated. The Boeing 702 bus uses deployable heat radiators to increase the radiating area. Without deployable radiators the dimensions of the bus limit the radiating area.

An alternative to deployable radiators is radiating TWTs. These TWTs have their collectors mounted outside the surface of the spacecraft where they radiate about half of their heat directly into space. Radiating TWTs have become commonplace in high-powered GSO satellites.

Inside the satellite thermostatically controlled heaters keep the temperature from falling below specifications as the heat dissipation of the payload varies. Power for the heaters is rarely a problem because they are needed primarily when the payload is not consuming its full power.

7.11.4 Tracking, Telemetry, and Control (TT&C)

The tracking subsystem is responsible for determining the orbital position of the satellite. The telemetry subsystem collects telemetry from all over the satellite and sends it to a ground control station. The control subsystem activates redundant units in the event of a failure, sets transponder gain and other payload parameters, and issues commands to correct for orbital drift. Solar radiation and uneven mass distribution of the earth pull a satellite out of its desired position. The TT&C subsystem links are used for ranging, to locate the position of a satellite, providing the information necessary to generate station-keeping commands to move the spacecraft to its desired orbital position.

The telemetry and command subsystems are usually completely separate subsystems with their own HPA and antennas. However in some cases payload capacity is used for telemetry and command. IRIDIUM uses the gateway section of the payload for the RF T&C link. If the mission (payload) antenna is used for TT&C, there must be a backup antenna with an omni radiation pattern that can be used to communicate with the satellite when it is not pointed to the earth, such as prior to its being in geosynchronous orbit or if earth lock is lost.

7.11.5 Attitude Control and Station Keeping

These subsystems work to keep the satellite in the correct orbit and to keep it accurately pointed toward the earth. This subsystem is quite complex. The important components are:

- Sun and earth sensors to tell the satellite where it is in relation to the earth
- Gyroscopes to provide real time updates of the satellite's orientation
- Thrusters to correct for spacecraft drift
- Momentum wheels to point the satellite
- Magnetic 'torquer' to change the momentum without using the thrusters by interacting with the external magnetic field
- Computers to process the sensor information and control the thrusters

The orbital maneuver life (OML) is determined by the amount of fuel on the spacecraft when it reaches its mission orbit divided by the amount of fuel consumed each year. Some fuel is usually reserved to move the satellite out of its orbit at the end of its mission.

7.12 SPACE SEGMENT TECHNOLOGY TRENDS

Advances in the spacecraft bus allow larger payloads by providing more power and more thermal dissipation. These enhancements are made possible by:

- More efficient solar cells allowing higher power payloads
- Deployable thermal radiators, allowing the increased power to be used by the payload, and
- Ion propulsion, reducing the mass required for station keeping fuel and allowing increased payload mass

Advances in digital and antenna technology allow major payload advances that increase payload capacity and increase link margins. Improved capacity and improved link margins are obtained by using smaller beams. This requires larger antennas with more complex feeds and also some way of achieving the desired connectivity. Advances in unfurlable antenna technology allow larger antennas with smaller spot beams. Digital technology provides the required connectivity and also makes entirely new concepts in phased array antennas possible.

Large unfurlable 12 meter diameter spacecraft antennas are already in use on Thuraya and ACeS, with 30 to 45 meter diameter antennas being developed. ACeS has 140 spot beams while Thuraya has 250 to 300 spot beams. With that many beams, digital processing but not necessarily regeneration is required to provide connectivity. Current digital technology allows a signal to be digitized, filtered, and switched, combined with other signals, and converted back to RF, but without regeneration. The signal is a digitized copy of the waveform. This allows small sections of spectrum to be routed, which is required when a large number of beams must be interconnected. Although the signal is digitized, this is considered a bent pipe satellite because the downlink signal is identical to the uplink signal except in frequency. It does not have the link advantages and as flexible routing of a regenerative payload but it offers more versatility in handling many types of signals, modulation formats, etc.

Digital technology permits narrow band commercial regenerative spacecraft such as IRIDIUM. As the computational power increases and power consumption decreases, regeneration will become more common in broadband systems for both link and routing advantages, especially where ISLs are used.

Advances in electronics are also making new architectures possible in spacecraft antenna design. Several commercial companies are actively pursuing digital beam forming (DBF) with Thuraya being the first commercial satellite to incorporate DBF. Each receive element in a DBF antenna digitizes the RF signal. Beam recovery and signal switching is done digitally. The signal may or may not be demodulated and is converted to RF at the transmit antenna. The RF phase and amplitude is controlled very accurately using digital techniques. A beam can be moved in any direction including tracking a stationary spot on the ground from a moving LEO satellite. This eliminates beam-to-beam handovers and the onboard switching resources otherwise required. It can allow each user to have a unique beam that tracks that user as he moves.

The advances in bus technology allow more power, mass and heat rejection for the payload. This will allow the new payload technologies to be incorporated into future spacecraft.

These techniques will allow links to PDAs, cell phones and high-speed data systems to any user anywhere in the world.

REFERENCES

BOOKS
Agrawal, Brij N. Design of Geosynchronous Spacecraft, Prentice-Hall Inc. 1986.

Feher, Kamilo, Digital Communications, Satellite / Earth Station Engineering, Prentice Hall, 1983.

Gordon, Gary D. and Morgan Walter L. Principles of Communications Satellites, John Wiley & Sons 1993.

Logsdon Tom, Mobile Communications Satellites, Theory and Applications, McGraw-Hill, 1995.

Maral G., and Bousquet M. Satellite Communications Systems, Third Edition, John Wiley & Sons, 1998.

Martin, James. Communication Satellite Systems, Prentice-Hall Inc. 1978.

Michelson Arnold M, Levesque Allen H., Error Control Techniques for Digital Communication, John Wiley and Sons, 1985.

Pattan Bruno, Satellite Systems Principles and Technologies, Chapman & Hall. 1993.

Pritchard, Wilbur L., Henri G. Suyderhoud, and Robert A. Nelson, Satellite Communication Systems Engineering, 2nd addition, Prentice-Hall Inc. 1993.

JOURNAL/SYMPOSIA ARTICLES
Neale, R Jason , Rod Green, and John Landovsis, "Interactive Channel for Multimedia Satellite Netwks," IEEE Communications Magazine., vol 39, no 3, March 2001 pp 192-198.

Wood, Lloyd, George Pavlou and Barry Evans, "Effects on TCP of Routing Strategies in Satellite Constellation," IEEE Communications Magazine, vol 39, no 3, March 2001 pp 182-181.

Yurong Hu and Victor O. K. Li, "Satellite-Based Internet: A Tutorial," IEEE Communications Magazine, vol39, no 3, Mar 2001 pp 154-162.

Chapter 8

GMSS Summary

Peter A. Swan[1] and Carrie L. Devieux Jr.[2]
[1] SouthWest Analytic Network, Inc,   [2] Chandler, Arizona

8.1   ACHIEVEMENTS

The remarkable achievement of the 1990's in the area of GMSS is that there are engineering marvels out there operating. The design was accomplished. The production techniques revolutionized the space arena. The launch campaigns were marvelous achievements. The operations concepts were definitely cutting edge in both the technologies of computers and software, but in the human capital as well. An amazing decade in engineering achievements. The systems are conducting operations as the designers had predicted. The vision was executed over the last twelve years.

Communications Anywhere, Anytime by Anyone

8.2   ARCHITECTURAL APPROACH

The development of mega-projects over centuries has taken men of vision and financiers that accepted risk. These achievements were necessary as the world progressed in the technological era of the 18th, 19th and 20th centuries. Each time the approach was essentially the same, yet not called architecture until many years later. First there was vision, then there were the tradeoffs for engineering design, user needs and artistic inputs. Each of these important components was traded within the scope of the program and resulted in major accomplishments. One of the key lessons learned over the years is that the proper approach, the proper motivation and the right ingredients do not always lead to the success expected. The Panama Canal was started multiple times, but finally was completed. The financial success of airplanes took decades to develop. Multiple railroads were started and had trouble maintaining a commercial success. Good business mixed with good engineering sometimes leads to excellent results. Often times, there are failures, slow-downs, and out right losses. However, the engineering process of starting with a vision and pushing through a mega-project with a diverse set of complex activities is a daunting task and deserves recognition by the user. Most times this does not happen. Often times, the question is what's next?

8.3   EVENTUAL SUCCESS

The eventual success of the GMSS concept will surface as time progresses. The idea of messaging from remote locations will ensure that the Little LEOs will find a niche market to provide a hefty profit. The need for roaming scientists, oil explorers, yatchmen, oil platform occupants, and "outback" workers will lead to the recognition that mobile phones are essential to safety, business success and comfort. The GSO orbit is an excellent location for the regional approach to the roaming caller. Each of these businesses will eventually surface as a success. The key is that the investors were assured that there was risk and that they should expect great returns for their money invested.... Or a great loss. This is the basis for capitalism. The opportunity to risk and gain much, or lose the investment. Many factors went into the financial difficulties of the GMSS players, not the least of which is the time lag from program approval (early 90's) to the start of revenue (late 90's). Even though the industry was amazed that so much engineering and production could be accomplished in such a short time, QUICKER is the word for the next generation. Five years from funding to revenue is too great a time lag when the business revolution is exploding around the globe and the technologies are racing away faster that orbiting satellites. Time to market is critical and will be a principle determinate for future generations of GMSS. Why can't the space industry execute a constellation of satellites for GMSS in the time period that the GSO suppliers provide a new satellite under a new contracts. How about two years to market for a constellation of satellites for a new business case? This expectation must be met!

8.4 THE FUTURE of GMSS

The future looks bright for the GMSS industry. There are assets in the space arena that are going to be utilized by new companies that "buy-out" old debt. These new companies will have the tremendous advantage of developing a business with minimum debt and maximum motivation to concentrate on the bottom line instead of the newest technological innovations. "Get on" with business is the new vision. Execution is the watch word.

The various types of Global Mobile Satellites Systems will have different niche markets. IRIDIUM and Globalstar have demonstrated their great value in many occasions such as disaster relief when other forms of communications have been rendered inoperative. The GSOs may be able to provide perhaps lower service cost since they require somewhat simpler less costly infrastructures. The great advantages of satellites are that they can provide ubiquitous coverage and supply badly needed communications where terrestrial systems are not yet well developed. Developing countries could use such capabilities to substantially improve their national economy, health services , educational systems and many other areas. One of the key challenges is of course, to find ways to bring the cost of handsets and services to a more affordable level.

8.5 THE CUSTOMER

The Customer will win in the long run. The GSO GMSS assets will provide regional coverage that will be focused on local issues while the Big LEOs will be focusing on roaming professionals in vertical markets. The Little LEOs will be charging ahead with their vision of reading meters from space, wherever they are. The future is now, the customer is king.

Index